THE ULTIMATE BOOK OF FACTS ABOUT EVERYTHING

Science, Animals, History, World Records, Inventions, Sports, Pop Culture, Technology, Space, Language, and More!

JACK HAYNES

ISBN: 978-1-957590-47-9

For questions, email: Support@AwesomeReads.org

Please consider writing a review!

Just visit: AwesomeReads.org/review

FREE BONUS

SCAN TO GET OUR NEXT BOOK FOR FREE!

TABLE OF CONTENTS

INTRODUCTION

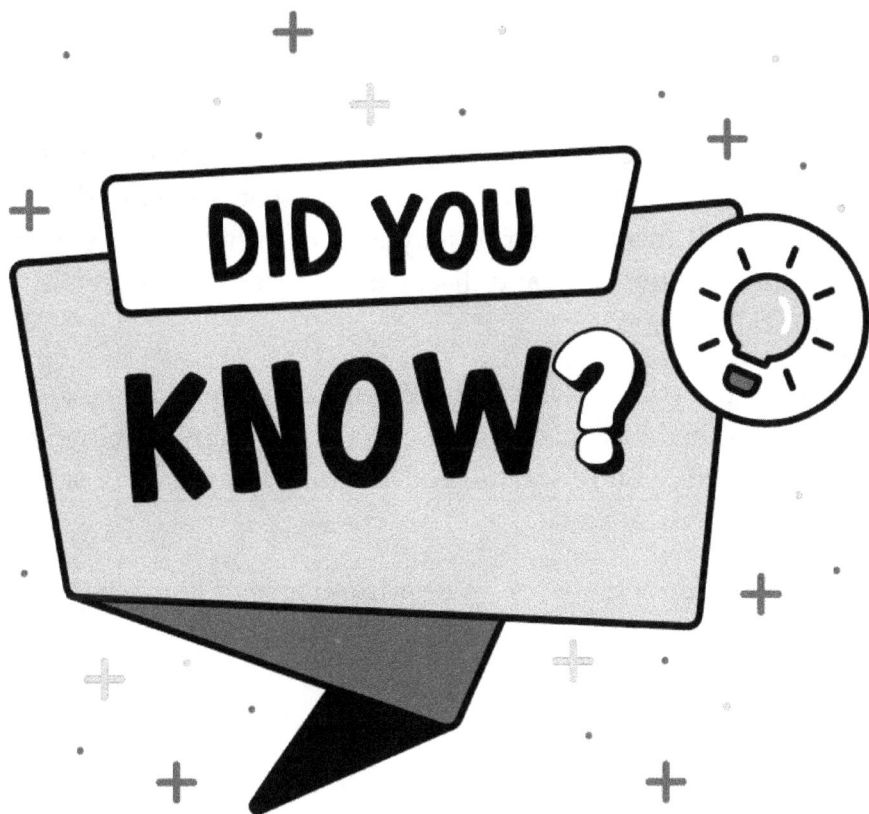

Curiosity drives us to explore and question the extraordinary world around us. This book celebrates that curiosity. It's a compendium of mind-bending science, fascinating history, and unexpected truths that's designed to both surprise and inform. Perfect for lifelong learners and trivia enthusiasts, this book invites you to dive headfirst into the facts that make reality feel stranger than fiction.

Each chapter will introduce you to a new topic, ranging from animals and space to pop culture and weird laws. This book explores the fascinating "why" behind what we see, do, and believe. You'll meet immortal jellyfish, learn how your stomach glows faintly in the dark, and discover forgotten heroes who saved the world by mistake.

Each section is packed with entertaining entries that can be read in any order. Whether you flip through a page at a time or binge read entire chapters, you're guaranteed to walk away a little smarter and a little more amazed.

So sharpen your wit, open your mind, and enjoy a journey through the world's delightful absurdities. Let's get curious.

CHAPTER ONE: SCIENCE

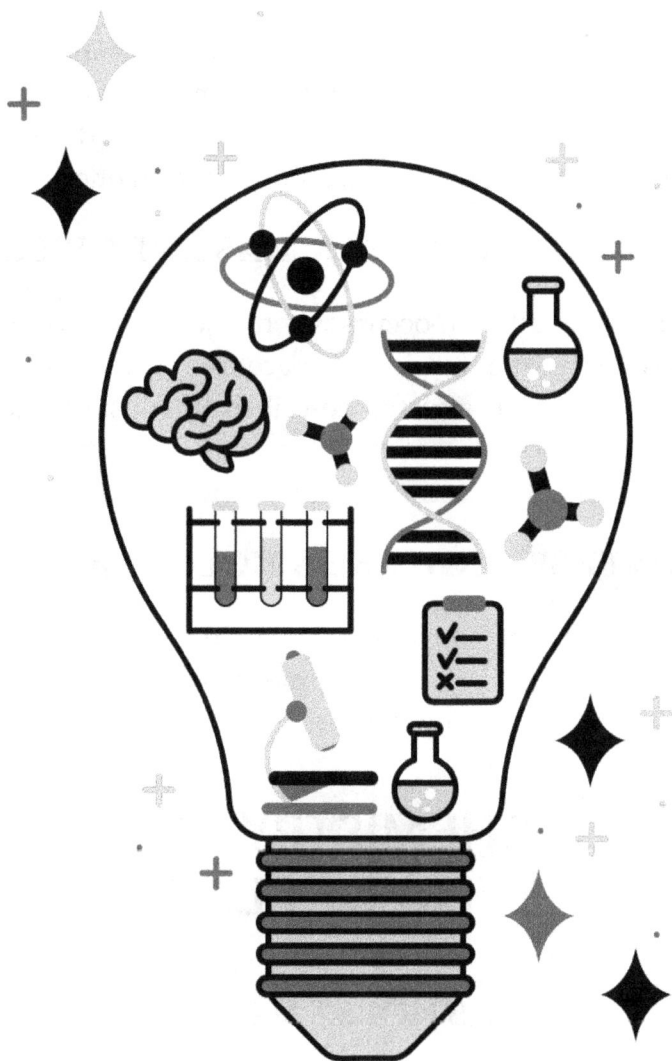

FASCINATING PHYSICS CONCEPTS

QUANTUM ENTANGLEMENT DEFIES DISTANCE

Famously called "spooky action at a distance" by Albert Einstein, quantum entanglement is a natural phenomenon where two or more particles are linked. In a nutshell, the particles behave identically no matter how far apart they are from one another.

TIME RUNS SLOWER NEAR MASSIVE OBJECTS

Gravitational time dilation occurs when objects such as planets or black holes warp the fabric of spacetime and increase their gravitational field. This effect is described by Einstein's theory of general relativity, which states that gravity isn't just a force but rather a curvature of spacetime that's affected by mass.

THE UNIVERSE MIGHT BE A HOLOGRAM

In 1993, Gerard 't Hooft proposed the holographic principle. His theory suggests that our three-dimensional reality is only a projection of two-dimensional information.

CHEMISTRY IN EVERYDAY LIFE

BAKING IS CHEMISTRY AT WORK

Chemical reactions are at the heart of transforming raw dough into baked goods. Agents such as baking soda or powder release carbon dioxide, which causes the baked goods to rise. Meanwhile, heat causes proteins and sugars to undergo the Maillard reaction during baking to create browning and flavor.

SOAP WORKS BECAUSE OF ITS MOLECULAR SHAPE

Soap molecules are unique because they have a hydrophilic (water-attracting) head and a hydrophobic (water-repelling) tail. This structure forms tiny spheres called micelles. Dirt and oil are trapped inside the micelles, which can be rinsed away easily with water.

YOUR BODY RUNS ON COMBUSTION

Cellular respiration is a controlled chemical reaction similar to combustion. With every breath, your cells combine glucose from food and oxygen to release energy, water, and carbon dioxide.

FRUITS CHANGE COLOR DUE TO PH

Pigments called anthocyanins change color depending on pH. For example, red cabbage turns pink in acids like lemon juice and green or blue when exposed to bases like baking soda.

BIOLOGY ODDITIES

TARDIGRADES CAN SURVIVE IN SPACE

More commonly known as water bears, tardigrades are known for their ability to slip into a near-death-like state called *cryptobiosis*. This process allows them to survive intense radiation, freezing, boiling temperatures, and extreme environments like the vacuum of outer space.

YOUR GUT HAS A "SECOND BRAIN"

The enteric nervous system is located in your digestive tract and contains about 500 million neurons! This neural network controls digestion independently of the brain and produces 90% of the body's serotonin, which helps regulate mood.

SOME ANIMALS ARE IMMORTAL

The *Turritopsis dohrnii* is colloquially called the "immortal jellyfish" on account of its ability to revert to its juvenile form after becoming an adult. This cycle can theoretically continue indefinitely, allowing it to escape death by aging. Despite this superpower, these jellyfish can still die from disease or injury.

YOU'RE MORE BACTERIA THAN HUMAN

Humans have about 30 trillion human cells and up to 39 trillion bacterial cells! Most of these microbes live in your gut and help with digestion, immunity, and even mental health.

ACCIDENTAL SCIENTIFIC BREAKTHROUGHS

PENICILLIN WAS DISCOVERED BY MOLD

Alexander Fleming accidentally discovered penicillin in 1928 after leaving a Petri dish of bacteria out while on vacation. When he returned, the mold growing on the dish was killing the bacteria. That mold became the world's first proper antibiotic.

X-RAYS WERE DISCOVERED WHILE PLAYING WITH ELECTRICITY

In 1895, Wilhelm Röntgen accidentally discovered X-rays while experimenting with cathode rays. He noticed that a nearby screen was glowing even though it was shielded. X-rays revolutionized medicine as a form of non-invasive imaging.

VULCANIZED RUBBER WAS A HAPPY MISTAKE

Charles Goodyear dropped a mixture of rubber and sulfur onto a hot stove. Instead of melting, it formed vulcanized rubber. This durable, flexible material is now used to create modern tires and waterproof boots.

PLASTIC WAS BORN FROM A BILLIARDS GAME

In 1869, a company offered a prize for finding a substitute for the ivory used in billiard balls. John Wesley Hyatt invented celluloid, the first synthetic plastic, while experimenting with different materials. This paved the way for the plastic revolution.

STRANGE EXPERIMENTS WITH UNEXPECTED RESULTS

THE STANFORD PRISON EXPERIMENT SPIRALED OUT OF CONTROL

In 1971, psychologist Philip Zimbardo set up a mock prison to study the psychological effects of authority. Volunteers were randomly assigned as guards or prisoners. Within days, the "guards" became abusive, and the "prisoners" showed signs of extreme stress. The experiment had to be shut down early, revealing how easily people conform to roles under pressure.

PAVLOV'S DOGS RESPONDED TO MORE THAN BELLS

Everyone knows Ivan Pavlov trained dogs to salivate at the sound of a bell, but it's less known that the dogs also began to react to lab coats, footsteps, and even the time of day. These unexpected associations showed how deeply conditioned responses could go beyond a single trigger.

THE MILGRAM EXPERIMENT SHOWED THAT PEOPLE OBEY ORDERS — EVEN TO THE HARM OF OTHERS

In the 1960s, Stanley Milgram tested how far people would go when ordered to give electric shocks to a stranger. Surprisingly, 65% of participants were willing to deliver what they believed were fatal shocks simply because an authority figure told them to do so. It exposed uncomfortable truths about obedience and morality.

SCHRÖDINGER'S CAT WAS MEANT TO BE ABSURD

The famous thought experiment by Erwin Schrödinger — that a cat can be both alive and dead until observed — was initially intended to criticize quantum mechanics, not explain it. Ironically, it became a symbol supporting the strange nature of quantum superposition.

EVOLUTIONARY QUIRKS IN NATURE AND HUMANS

HUMANS HAVE A TAILBONE BUT NO TAIL

The coccyx, or tailbone, is a leftover from our primate ancestors with fully functional tails. Although we no longer need tails for balance or climbing, the bone still remains. It's a classic example of a vestigial structure in evolution.

WHALES HAVE LEG BONES

Modern whales evolved from land-dwelling mammals. Deep inside their bodies, they still have tiny leg bones—remnants of their ancestors who once walked on land over 50 million years ago.

THE LARYNGEAL NERVE TAKES A RIDICULOUS DETOUR

In giraffes (and humans), the recurrent laryngeal nerve runs from the brain to the voice box but detours all the way down the neck and back up again. That's about 15 feet of unnecessary looping in giraffes, all because of how this nerve was "rewired" through evolution.

GOOSEBUMPS ARE AN ANCIENT REFLEX

When you're cold or scared, your skin gets goosebumps as tiny muscles at the base of your hair follicles contract. This reflex was useful when our ancestors had more body hair; it would puff up to trap heat or make them look bigger.

SCIENTIFIC PARADOXES

THE TWIN PARADOX (TIME TRAVEL BY SPEED)

In Einstein's theory of relativity, if one twin travels near the speed of light and comes back, they'll have aged less than the twin who stayed on Earth. It's not science fiction; it's been tested with atomic clocks on airplanes. Time literally slows down at high speeds!

THE BOOTSTRAP PARADOX (TIME LOOP TROUBLE)

This time travel paradox asks: What if you go back in time and give Shakespeare a copy of *Hamlet* and he publishes it as his own work? Who actually wrote it? The origin of the information is lost in an endless loop that defies logic.

THE FERMI PARADOX (WHERE ARE THE ALIENS?)

Given the vast number of planets in the universe, intelligent alien life should statistically exist and we should've encountered it by now. But we haven't. So where is everybody? The Fermi Paradox makes us question our assumptions about life and the cosmos.

THE MONTY HALL PROBLEM (PROBABILITY TRICKERY)

You're on a game show with three doors. One has a car; the others have goats. You pick one. The host opens another to reveal a goat. Should you switch? Yes—always. Switching gives you a two-thirds chance of winning, but most people intuitively stick with their original choice. This paradox messes with our understanding of probability.

FAMOUS SCIENTISTS AND THEIR LESSER-KNOWN DISCOVERIES

ISAAC NEWTON WAS AN ALCHEMIST

Most people know Newton for his work on gravity and calculus. However, he also spent decades studying alchemy while trying to turn base metals into gold and uncover the "Philosopher's Stone." He even wrote more on alchemy and biblical prophecy than physics.

MARIE CURIE DISCOVERED MORE THAN JUST RADIUM

Marie Curie is best known for discovering radium and polonium, but she also developed mobile X-ray units during World War I. These "Little Curies" helped over a million soldiers and revolutionized battlefield medicine.

NIKOLA TESLA HAD WIRELESS INTERNET-LIKE IDEAS

While Tesla is famous for AC electricity, he also envisioned a global wireless communication system in the early 1900s. He built the Wardenclyffe Tower to send wireless messages—and even wireless power—worldwide, long before Wi-Fi existed.

ALBERT EINSTEIN INVENTED A FRIDGE

Einstein, concerned about toxic gas leaks in early refrigerators, co-invented an eco-friendly absorption refrigerator with no moving parts in 1926. Even though it wasn't mass produced, the concept is still studied for developing sustainable cooling systems.

CHAPTER TWO: ANIMALS

EXTREME ANIMAL ABILITIES

PISTOL SHRIMP'S SONIC WEAPON

The pistol shrimp snaps its claw so fast that the resulting bubble briefly reaches temperatures of over 8,500°F. That's nearly as hot as the sun's surface. This cavitation bubble stuns or kills prey and produces a sound loud enough to interfere with sonar equipment.

FLEAS' JUMPING MIGHT

Fleas can jump over 100 times their body length by exerting a force of over 100 times gravity. They store energy in a special protein called resilin that acts like a biological spring to catapult themselves into the air.

AXOLOTL: THE REGENERATION CHAMPION

The axolotl, a type of salamander, can regrow its limbs, spine, heart, eyes, and even parts of its brain without scarring. Scientists study it closely in hopes of unlocking human regenerative potential.

BATS: ACOUSTIC NAVIGATORS OF THE NIGHT

Most bats use echolocation by emitting high-frequency sounds and analyzing the echoes that bounce off objects. This allows them to hunt flying insects in complete darkness with incredible accuracy. They can detect prey as fine as a human hair in midair.

UNIQUE MATING RITUALS IN THE WILD

THE PUFFERFISH'S UNDERWATER CROP CIRCLES

Male Japanese pufferfish create elaborate geometric sand patterns on the ocean floor — up to seven feet wide — to attract a mate. These intricate "crop circles" are sculpted with fins over days and serve no other purpose than impressing females.

THE UNUSUAL ROMANCE OF THE ANGLERFISH

Male anglerfish are tiny compared to females — sometimes one-tenth the size. During mating, the male bites and fuses to the female's body. The male slowly dissolves until only his testes remain, permanently supplying sperm while the female continues hunting.

BOWERBIRDS: INTERIOR DESIGNERS OF THE BIRD WORLD

Male bowerbirds build elaborate structures made of twigs and meticulously decorate them with colorful objects. Blue is especially popular. Females judge mates based on artistic flair and tidiness.

REDBACK SPIDERS: SACRIFICIAL MATING

Male redback spiders (a species of black widow) often sacrifice themselves during mating by somersaulting into the female's fangs. This increases the chances of successful fertilization. Furthermore, she may be less likely to mate again, giving his genes a better shot at surviving.

ANIMALS WITH UNEXPECTED INTELLIGENCE

CROWS UNDERSTAND CAUSE AND EFFECT

Crows can solve multi-step puzzles, use tools, and even understand water displacement. In experiments, they've dropped stones into tubes of water to raise the water level and reach floating food.

PIGS ARE AS SMART AS DOGS

Pigs can play video games with joysticks, recognize their own reflections (a sign of self-awareness), and remember complex mazes. They also form social hierarchies and can deceive other pigs to protect their food sources.

OCTOPUSES SOLVE PUZZLES AND ESCAPE ROOMS

Octopuses have incredible problem-solving abilities. They've been known to unscrew jar lids, mimic other species, and escape tanks through tiny holes. Their intelligence is decentralized, with two-thirds of their neurons in their arms.

GOATS CAN READ HUMAN EMOTION

Goats aren't just clever climbers. They can interpret human facial expressions and prefer to approach smiling faces. Studies show they can also remember solutions to puzzles for months after learning them.

BIZARRE DEFENSE MECHANISMS

LIZARDS SHOOT BLOOD FROM THEIR EYES

Texas horned lizards can shoot blood out of their eyes (up to five feet away) to startle predators. The blood contains chemicals that taste foul to canines like coyotes, helping the lizard escape.

SEA CUCUMBERS EJECT THEIR GUTS

When threatened, sea cucumbers literally vomit their internal organs—sometimes including their intestines—to entangle or distract predators. They can regenerate everything they lose in a matter of weeks.

HAIRY FROGS BREAK THEIR OWN BONES

Also known as the "Wolverine frog," this species can break bones in its toes, which then pierce through the skin to form claws. Scientists believe this painful move is used as a last-ditch defense when attacked.

BOMBARDIER BEETLES SPRAY CHEMICALS

These beetles mix chemicals inside their abdomens to create a scalding hot chemical spray that reaches nearly 212°. They can aim this explosive burst in almost any direction to fend off predators.

OPOSSUMS ARE MASTERS OF PLAYING DEAD

Opossums don't just faint when threatened. They go into a full, involuntary catatonic state, complete with tongue lolling, drooling, and a corpse-like stench. The smell comes from a gland that produces a foul fluid to deter would-be predators.

ANIMAL RECORD HOLDERS

PEREGRINE FALCON: FASTEST ANIMAL ON EARTH

The peregrine falcon holds the record for the fastest dive speed of any animal. It can reach speeds over 240 mph when diving to catch prey, making it faster than a Formula 1 car!

WATER BOATMAN: LOUDEST ANIMAL RELATIVE TO SIZE

The tiny water boatman (specifically *Micronecta scholtzi*) is just 2 mm long but can produce sounds up to 99 decibels, or as loud as a jackhammer. It does this by rubbing its penis against its abdomen in a process called stridulation.

ARCTIC TERN: LONGEST MIGRATION

The Arctic tern migrates from the Arctic to Antarctica and back every year, covering over 44,000 miles annually. That's the longest known migration route of any animal on the planet.

MING THE CLAM: OLDEST KNOWN ANIMAL

A deep-sea ocean quahog clam nicknamed Ming was found to be 507 years old. Born in 1499, it lived through the entire Renaissance and was accidentally killed by researchers during a study.

DUNG BEETLE: STRONGEST ANIMAL

The horned dung beetle can pull over 1,100 times its body weight. That's like a human dragging six double-decker buses. Pound for pound, it's the strongest creature on Earth.

18

ENDANGERED AND EXTINCT SPECIES WITH STRANGE HISTORIES

THE BAIJI DOLPHIN: EXTINCT FROM SILENCE

The baiji, or Yangtze River dolphin, was declared functionally extinct in 2006. Its extinction happened quietly in one of the world's longest and most culturally significant rivers without many people even noticing. It's considered the first aquatic mammal extinction caused by humans.

THE LAZARUS LIZARD: RESURRECTED FROM EXTINCTION

The La Palma giant lizard (*Gallotia auaritae*) was thought to be extinct for 500 years. It was only known from fossil remains until a living specimen was found in 2007. It was named after Lazarus, who rose from the dead, sparking hope for other "extinct" species.

THE PASSENGER PIGEON: FROM BILLIONS TO ZERO IN 50 YEARS

In the early 1800s, passenger pigeons were the most abundant bird in North America; flocks could darken the sky for hours. Yet by 1914, they were completely extinct due to commercial hunting and deforestation. Their story is a chilling example of how fast abundance can vanish.

THE PYRENEAN IBEX: CLONED AND EXTINCT AGAIN

The Pyrenean ibex went extinct in 2000. In 2003, scientists used DNA to clone a new individual that died shortly after birth. It became the first species to go extinct, be brought back to life, and go extinct again.

ANIMALS WITH HUMAN-LIKE BEHAVIOR

ELEPHANTS HOLD FUNERALS

Elephants are known to grieve by gently touching the bones of lost herd members with their trunks. They sometimes stand vigil over a dead companion and have even been observed trying to bury or cover the body with leaves and branches.

CAPUCHIN MONKEYS USE TOOLS

Capuchin monkeys use rocks to crack open nuts, but they also display complex emotions. In a famous experiment, they refused to perform a task if they saw a neighbor being rewarded with a better treat, essentially showing envy and a sense of fairness.

PRAIRIE DOGS HAVE A COMPLEX LANGUAGE

Prairie dogs have one of the most complex communication systems ever discovered in animals. Their calls can describe types of predators, size, shape, speed, and even the clothing color of approaching humans. It's the closest thing we've found to a grammatical animal language.

BONOBOS USE EMPATHY TO SOLVE CONFLICT

Bonobos, one of our closest primate relatives, often use sex and physical affection to defuse tension and form bonds. Their emotional intelligence and social strategies make them remarkably similar to humans.

CREATURES FOUND IN EXTREME ENVIRONMENTS

ICEFISH HAVE ANTIFREEZE IN THEIR BLOOD

Living in the freezing waters of Antarctica, icefish have evolved a special type of antifreeze glycoprotein in their blood that prevents their bodily fluids from freezing. Their blood is also clear, as they lack red blood cells entirely!

TUBE WORMS THRIVE IN BOILING HYDROTHERMAL VENTS

Giant tube worms (*Riftia pachyptila*) live over a mile underwater near hydrothermal vents where temperatures can exceed 750°F. They have no mouths or stomachs. Bacteria living inside them convert toxic vent chemicals into energy in a process called chemosynthesis.

DEVIL WORMS LIVE MILES UNDERGROUND

The *Halicephalobus mephisto*, nicknamed the "devil worm," is the deepest-living multicellular organism ever discovered. It thrives 2.2 miles beneath the Earth's surface where temperatures reach 131°F and oxygen is scarce.

SAHARA SILVER ANTS RACE THE HEAT

The Sahara silver ant is one of the fastest ants on Earth, running at speeds of up to 108 body lengths per second. It forages during the hottest part of the day when surface temperatures exceed 140°F to avoid predators that can't take the heat.

LAKE NATRON FLAMINGOS NEST ON CAUSTIC SODA FLATS

Lesser flamingos breed on the shores of Lake Natron in Tanzania, a lake so alkaline that it can burn skin and eyes. With a pH up to 10.5, it can even preserve animals like stone statues. Despite the harsh conditions, flamingos nest in the salty crusts where they're protected from predators that can't handle the lake's toxicity.

CHAPTER THREE:
THE HUMAN BODY

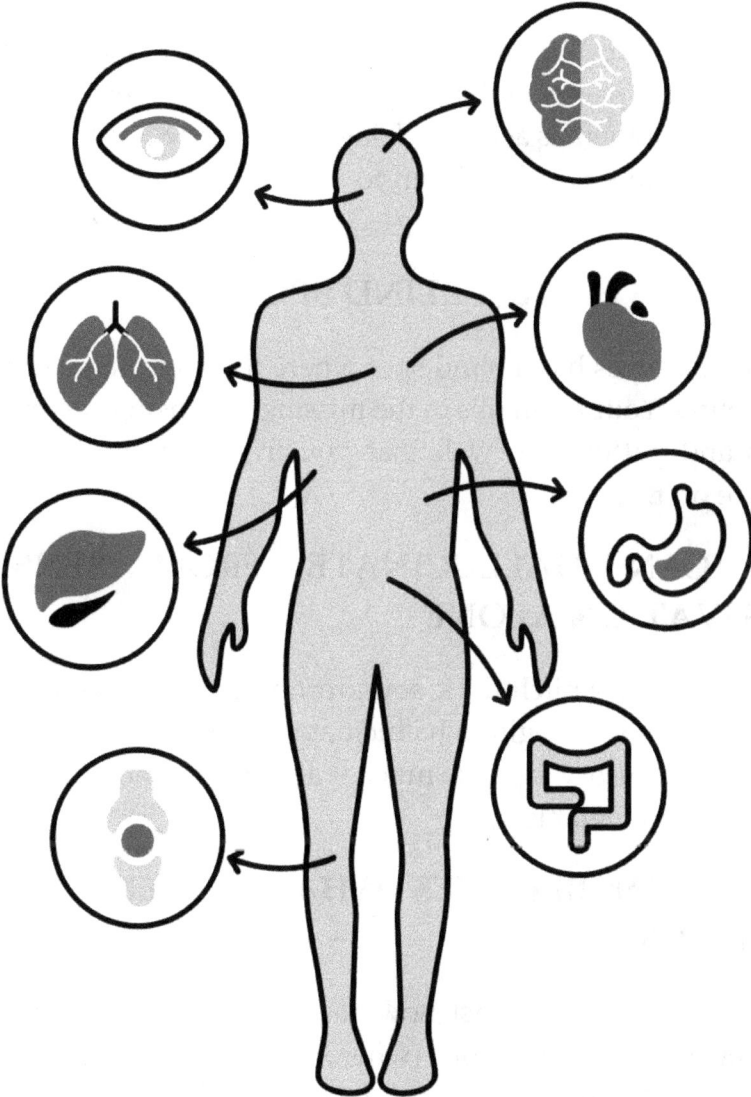

BRAIN POWER AND PERCEPTION TRICKS

YOUR BRAIN EDITS REALITY

Your brain takes in fragmented sensory data and stitches it together into a seamless "now." However, what you perceive as the present moment is actually delayed by about 80 milliseconds — just enough time for your brain to smooth over errors and fill in blanks.

YOUR EYES HAVE A BLIND SPOT

Each of your eyes has a blind spot where the optic nerve connects to the retina. Your brain fills in the missing data using surrounding details and patterns so well that you live your life completely unaware of it.

YOU CAN HALLUCINATE FROM SENSORY DEPRIVATION ALONE

Lock someone in a pitch-dark, soundproof room and even the most rational minds will begin to hallucinate after just a few hours. The brain craves sensory input so much that it starts generating its own when none is available.

THE BRAIN BELIEVES WHAT IT SEES... EVEN WHEN IT'S WRONG

Optical illusions reveal just how easily fooled we are. From the "checker shadow" illusion (where two squares that appear different are the exact same color) to the Rubin vase (faces or a vase?), these tricks show that your brain only interprets reality.

STRANGE BODILY FUNCTIONS

YOU GLOW IN THE DARK (JUST VERY FAINTLY)

Humans emit a tiny amount of visible light due to biochemical reactions called bioluminescence. It's about 1,000 times weaker than the human eye can detect, but high-sensitivity cameras have captured this ghostly glow.

YOU CAN SMELL MEMORIES — LITERALLY

Your sense of smell is directly connected to the brain's emotional center. That's why a single whiff of something like sunscreen or cookies can instantly trigger vivid, emotional memories faster than any other sense.

YOUR BONES ARE STRONGER THAN CONCRETE

Ounce for ounce, human bone is about five times stronger than concrete. The femur (thigh bone) is so dense and resilient that it can withstand thousands of pounds of pressure, making it one of the toughest materials in your body.

YOUR BRAIN CAN "SEE" WITH YOUR TONGUE

A device called BrainPort converts visual information into tiny electrical impulses on the tongue. Amazingly, blind users have learned to interpret the patterns as shapes and movement, essentially "seeing" with their tongues.

HOW THE FIVE SENSES
WORK IN SURPRISING WAYS

YOU CAN "HEAR" SILENT GIFS

Many people experience a phenomenon called the "visual ear," where watching a silent video (like a bouncing power line or crashing object) triggers the illusion of hearing a sound. This is due to cross-wiring between visual and auditory processing centers.

YOUR NOSE CAN DETECT OVER A TRILLION SCENTS

It was once believed that humans could only detect around 10,000 smells. New research shows that the real number is closer to a trillion, far beyond what most animals can manage. Your olfactory system is more powerful than previously imagined, especially when combined with memory and emotion.

YOUR SKIN HAS MULTIPLE TOUCH MAPS

Touch isn't just one sense; it combines pressure, temperature, texture, pain, and vibration receptors. Some areas, like the lips and fingertips, have incredibly high nerve density that allows you to detect even the slightest changes.

YOUR EYES SEE EVERYTHING UPSIDE DOWN

The lens of your eye flips the world upside down, but your brain automatically inverts the image. In experiments where people wore special goggles that reversed the world, the brain adjusted after a few days, showing how plastic and adaptable the senses are.

MEDICAL ODDITIES AND RARE CONDITIONS

STONE MAN SYNDROME TURNS MUSCLE TO BONE

With fibrodysplasia ossificans progressiva, the body's soft tissues gradually turn to bone, essentially freezing the person alive in a second skeleton. There is no cure, and even minor injuries can trigger bone growth.

FOREIGN ACCENT SYNDROME IS REAL

After a brain injury or stroke, some people develop new accents they've never spoken before, despite never visiting those countries. This is a rare speech disorder caused by changes in brain function, not mimicry.

PEOPLE CAN HAVE EXTRA ORGANS OR LIMBS

Rare conditions like polymelia can result in a person being born with extra limbs, while others may have duplicated internal organs like two hearts or two uteruses. In most cases, these conditions are nonfunctional but biologically fascinating.

SOME PEOPLE CAN REMEMBER EVERY DAY OF THEIR LIVES

Highly superior autobiographical memory allows individuals to recall nearly every day of their lives in vivid detail, including what they wore or heard on the news. It's extremely rare and not fully understood, but people with the condition describe it as both a gift and a burden.

MYTHS ABOUT THE HUMAN BODY DEBUNKED

YOU USE MORE THAN 10% OF YOUR BRAIN

This myth has been completely debunked. Brain imaging shows that nearly every part of the brain is active at different times, even when you're sleeping. You use 100% of your brain, just not all at once.

HAIR AND FINGERNAILS DON'T KEEP GROWING AFTER DEATH

It only looks that way. After death, the skin dehydrates and shrinks, exposing more of the nails and hair. Actual growth stops completely because cells are no longer functioning.

SWALLOWED GUM DOESN'T STAY IN YOUR STOMACH FOR SEVEN YEARS

While gum isn't digested like other foods, it passes through your system like anything else, usually within a few days. Your intestines are well equipped to handle it unless you swallow massive amounts, which could cause blockages.

CRACKING YOUR KNUCKLES DOESN'T CAUSE ARTHRITIS

Multiple studies have shown that knuckle cracking doesn't increase your risk of arthritis. The cracking sound is caused by gas bubbles collapsing in the synovial fluid between your joints, not bones grinding together.

HOW THE BODY ADAPTS
TO EXTREME CONDITIONS

SOME PEOPLE CAN HOLD THEIR BREATH FOR OVER 10 MINUTES

Elite freedivers and the Bajau people of Southeast Asia have developed the ability to hold their breath underwater for incredible lengths of time. Studies show that the Bajau even have enlarged spleens, which help release more oxygen-rich blood during dives.

YOUR BODY CAN HIBERNATE IN COLD WATER

In freezing temperatures, especially when immersed in near-zero water, the body can activate a diving reflex that slows the heart rate, redirects blood to vital organs, and reduces oxygen use. This ability has sometimes allowed people to survive drowning for over an hour.

YOU CAN TRAIN TO RESIST EXTREME HEAT

People living in desert climates or practicing heat training in saunas or ultramarathons undergo physiological changes. They experience increased sweat efficiency, better salt regulation, and a lower core temperature during exertion. Your body literally learns to cool itself faster.

ZERO GRAVITY WEAKENS BONES AND MUSCLES, BUT THE BRAIN FIGHTS BACK

In space, astronauts lose muscle mass and bone density due to the lack of gravity. However, the brain compensates by rewiring balance systems and enhancing visual and touch sensitivity. With exercise and technology, astronauts now maintain much of their strength even in orbit.

SUPERHUMAN CAPABILITIES

SOME PEOPLE DON'T FEEL PAIN

People with a rare condition called congenital insensitivity to pain don't feel physical pain. While this may sound like a superpower, it's actually dangerous. These individuals can break bones or suffer burns without realizing it. It's due to mutations in a gene that blocks pain signals.

A REAL-LIFE IRON MAN EXISTS

Athlete Wim Hof, also known as "the Iceman," can withstand freezing temperatures for hours. He's run marathons in the Arctic barefoot, climbed snowy mountains in shorts, and submerged himself in ice baths. It's all thanks to his unique control over breathing, circulation, and stress response.

PEOPLE WITH EIDETIC MEMORY CAN RECALL IMAGES IN PERFECT DETAIL

While true photographic memory is rare, some people—especially certain children—have eidetic memory. This skill allows them to recall images or sounds in startling detail, almost like rewinding a mental video.

CHAPTER FOUR: HISTORY

FORGOTTEN MOMENTS
THAT CHANGED HISTORY

A CLERICAL ERROR HELPED END THE COLD WAR

In 1983, Soviet officer Stanislav Petrov received warning of an incoming U.S. nuclear strike. Instead of retaliating, he judged it to be a false alarm and trusted instinct over protocol. He was right. His calm decision averted global nuclear war and went largely unrecognized for decades.

A VOLCANO TRIGGERED THE BIRTH OF GOTHIC FICTION

In 1815, Mount Tambora erupted in Indonesia, causing the "Year Without a Summer." Crops failed, skies darkened, and a rainy vacation trapped Mary Shelley indoors with friends. The result? She wrote *Frankenstein*, helping to launch modern science fiction and horror.

A "GHOST ARMY" FOOLED HITLER'S FORCES

During WWII, the U.S. created a secret unit called the Ghost Army, which was made up of artists, sound engineers, and designers. They used inflatable tanks and fake radio chatter to mislead German troops. The ruse saved thousands of lives with illusion and creativity.

HIDDEN STORIES BEHIND FAMOUS EVENTS

PAUL REVERE WASN'T THE ONLY MIDNIGHT RIDER

While Paul Revere gets the spotlight, two other men — William Dawes and Samuel Prescott — also rode to warn colonists of British troops in 1775. In fact, Prescott was the only one to make it all the way to Concord. Revere was briefly captured.

THE STATUE OF LIBERTY WAS ORIGINALLY A LIGHTHOUSE

After being gifted by France, the Statue of Liberty was outfitted with a beacon to function as a lighthouse in 1886. However, it was too dim to be effective, and the plan failed.

CHURCHILL'S FAMOUS SPEECH WASN'T BROADCAST LIVE

One of history's most famous speeches — Winston Churchill's "We shall fight on the beaches" — wasn't heard live by the public. It was delivered to Parliament in 1940, and Churchill later recorded a version for radio. Most people heard a reenactment, not the original.

ECCENTRIC
LEADERS AND RULERS

EMPEROR NERO COMPETED IN THE OLYMPICS AND MADE SURE HE WON

Roman Emperor Nero insisted on competing in the Olympic Games (which were normally closed to non-Greeks). He entered a chariot race with 10 horses (twice the norm), fell off mid-race, and still declared himself the victor. The judges, fearing his wrath, agreed.

TSAR PETER THE GREAT TAXED BEARDS

Obsessed with modernizing Russia, Peter the Great imposed a beard tax in 1698 to make Russian men look more European. If you wanted to keep your beard, you had to pay and carry a token proving you'd paid. The tokens were engraved with a warning: *The beard is a useless burden.*

KING LUDWIG II OF BAVARIA BUILT A FAIRY-TALE WORLD

Ludwig II, obsessed with fantasy and opera, spent his reign building elaborate castles—most famously Neuschwanstein, which later inspired the castle in *Sleeping Beauty*. He lived alone in the fantasy world he created and died mysteriously in a lake at age 40.

CALIGULA MADE HIS HORSE A PRIEST

The notoriously unhinged Roman Emperor Caligula loved his horse, Incitatus. Besides giving him a marble stall and a jeweled collar, he appointed him a priest and reportedly planned to make him a senator. Some historians say it was political mockery; others say Caligula had gone mad.

EVERYDAY LIFE IN ANCIENT CIVILIZATIONS

ANCIENT EGYPTIANS WORE PERFUMED WAX CONES ON THEIR HEADS

In hot weather and during celebrations, Egyptians wore scented wax cones on their heads. As the day warmed up, the cones would melt slowly, releasing perfume over their bodies. It was a practical form of ancient deodorant and aromatherapy.

INCAS USED KNOTS INSTEAD OF WRITING

The Inca had no written language. Instead, they used strings with intricate knots and colors to record numbers, census data, and history. These "talking knots" could be read by trained record-keepers and were vital to running the vast Incan empire.

VIKINGS WERE SURPRISINGLY CLEAN

Contrary to the dirty barbarian stereotype, Vikings bathed at least once a week and used combs, tweezers, and ear picks. They even bleached their hair with lye soap. Archaeologists have found many grooming tools in Viking graves.

ANCIENT CHINESE USED "NIGHT SOIL" AS FERTILIZER

Farmers in ancient China collected human waste, or "night soil," from chamber pots and outhouses to fertilize crops. This early example of waste recycling was surprisingly effective.

BIZARRE HISTORICAL BELIEFS AND MEDICAL PRACTICES

DOCTORS ONCE PRESCRIBED TOBACCO TO TREAT ASTHMA

In the 17th and 18th centuries, tobacco was considered a medicinal herb. Doctors recommended smoking it to clear the lungs and ease asthma. Some even used tobacco smoke enemas to "revive" drowning victims, a practice promoted by respected medical societies!

VICTORIANS FEARED WANDERING WOMBS

In ancient Greece and even into the 19th century, some believed that a woman's uterus could roam around her body, causing hysteria and emotional instability. Treatments ranged from perfume near the vagina to sweet drinks, all to "lure" the womb back into place.

DEAD MOUSE PASTE FOR TOOTHPASTE

In Elizabethan England, one cure for a toothache involved crushing a dead mouse and applying it directly to the sore tooth. Other treatments included onion poultices or bleeding the gums with leeches.

MERCURY WAS A POPULAR MEDICINE

For centuries, the toxic metal mercury was used as a treatment for everything from syphilis to skin disorders. Even emperors like China's Qin Shi Huang and European nobles took mercury elixirs, believing it granted immortality. Ironically, it often led to madness and death.

TIMELINE OVERLAPS THAT SURPRISE PEOPLE

THE LAST U.S. CIVIL WAR WIDOW LIVED TO SEE THE INVENTION OF THE IPHONE

The last verified U.S. Civil War veteran, Albert Woolson, died in 1956. But the last widow of a Civil War veteran, Maudie Hopkins, lived until 2008, the same year Barack Obama was elected and the iPhone 3G was released.

NINTENDO WAS FOUNDED THE SAME YEAR JACK THE RIPPER ROAMED LONDON

Nintendo, the video game giant, originally started as a playing card company. It was founded in 1889, the same year Jack the Ripper was terrorizing London. In fact, Nintendo is older than plastic, airplanes, and sliced bread.

OXFORD UNIVERSITY IS OLDER THAN THE AZTEC EMPIRE

Teaching began at Oxford around 1096. That's over 300 years before the Aztec Empire was founded in 1325. Scholars were debating Aristotle in England while Tenochtitlán was still a swamp.

THE GUILLOTINE WAS STILL IN USE WHEN STAR WARS CAME OUT

France last used the guillotine for execution in 1977, the same year *A New Hope* premiered. That means lightsabers and public beheadings were, for a moment, part of the same world.

MILITARY MISTAKES THAT LED TO MAJOR CHANGES

NAPOLEON INVADED RUSSIA IN THE WINTER

In 1812, Napoleon launched a massive invasion of Russia but failed to account for the brutal Russian winter and the enemy's scorched-earth tactics. Of the 600,000 troops he began with, only about 100,000 survived. The disaster weakened Napoleon's empire and set the stage for his eventual defeat.

THE CHARGE OF THE LIGHT BRIGADE WAS A MISUNDERSTOOD ORDER

During the Crimean War in 1854, British cavalry was ordered to charge the wrong position due to vague communication and poor reconnaissance. The Light Brigade charged directly into a heavily fortified Russian artillery unit, leading to huge casualties and inspiring one of the most famous war poems in English history.

HITLER DELAYED AT DUNKIRK AND LET 300,000 TROOPS ESCAPE

In 1940, Hitler ordered a halt to German armored divisions outside Dunkirk, believing the Luftwaffe could finish the job. That pause allowed over 330,000 British and French troops to evacuate. It saved the core of the British army and changed the course of WWII.

PEARL HARBOR OVERSIGHT IGNORED RADAR WARNINGS

On the morning of December 7, 1941, a U.S. radar station detected incoming planes, but it dismissed them as a group of American bombers due in from the mainland. Hours later, Pearl Harbor was attacked, pulling the U.S. into WWII—a war it had previously resisted entering.

THE MAGINOT LINE WAS BUILT TO STOP GERMANY BUT FACED THE WRONG WAY

France built the Maginot Line, an enormous system of fortifications to stop German invasion. The Germans simply went around it and invaded through Belgium in 1940. The costly, immobile defense line became a symbol of fighting the last war, not the next one.

ORIGINS OF HISTORICAL MYTHS AND LEGENDS

KING ARTHUR MAY HAVE BEEN BASED ON MULTIPLE REAL PEOPLE

The legendary King Arthur is likely a composite of several Celtic warlords, not a single medieval king. The earliest mentions come from sixth-century Welsh poetry, but the Round Table, Excalibur, and Merlin were added centuries later by romantic storytellers.

VIKINGS NEVER WORE HORNED HELMETS

Despite their iconic image, no archaeological evidence supports Vikings wearing horned helmets. The myth likely began in the 19th century, thanks to the costume designers for Wagner's operas. Real Viking helmets were practical and horn free.

THE TROJAN HORSE MAY HAVE BEEN A METAPHOR OR AN EARTHQUAKE

The famous wooden horse from Homer's Iliad might never have existed. Some scholars believe it was a poetic metaphor for a siege engine or possibly a mistranslation of *ship*. Others suggest that Troy may have fallen due to an earthquake or fire, not Greek trickery.

THE SALEM WITCH TRIALS WERE FUELED BY MOLDY BREAD

One theory behind the hysteria of 1692 is that townspeople ingested ergot, a toxic mold found in rye bread that causes hallucinations and seizures. While controversial, it shows how environmental factors may have played a role in the infamous witch trials.

CHAPTER FIVE:
WORLD RECORDS AND EXTREMES

A MAN RAN 350 MILES WITHOUT SLEEP

Ultra-endurance athlete Dean Karnazes ran 350 miles in 80 hours and 44 minutes—without sleeping—during a solo run across California. His body recycles lactic acid so efficiently that he can keep running long after most people collapse from exhaustion.

IT'S POSSIBLE TO HOLD YOUR BREATH UNDERWATER FOR 24 MINUTES

In 2021, freediver Budimir Šobat held his breath underwater for an astonishing 24 minutes and 37 seconds. His extreme control of heart rate and metabolism, plus hyperventilation with pure oxygen, allowed him to push the limit to near-superhuman levels.

BLIND PEOPLE CAN LEARN TO ECHOLOCATE LIKE BATS

Some blind individuals, like Daniel Kish, have trained themselves to use clicking sounds and echoes to navigate the world. Their brains rewire to interpret sound spatially. It's a form of echolocation that's similar to techniques used by dolphins and bats.

THE HUMAN BODY CAN ADAPT TO LIFE AT THE "DEATH ZONE"

Sherpas and high-altitude climbers regularly survive in the "Death Zone" above 26,000 feet where oxygen is too scarce for most humans to function. Their bodies adapt by producing more red blood cells and increasing capillary density, allowing them to endure what would kill others.

TALLEST, LONGEST, OLDEST, AND FASTEST THINGS

TALLEST LIVING THING: THE HYPERION TREE

The tallest living tree on Earth is Hyperion, a coast redwood in California's Redwood National Park. It stands at 379.7 feet, making it taller than the Statue of Liberty and Big Ben. Its exact location is kept secret to protect it.

OLDEST KNOWN LIVING ORGANISM: A 9,500-YEAR-OLD SPRUCE

A Swedish clonal tree system called Old Tjikko has been growing from the same root system for over 9,500 years. That's older than all written human history.

LONGEST CONTINUOUS MOUNTAIN RANGE: THE MID-OCEAN RIDGE

The Mid-Ocean Ridge is an underwater mountain range that stretches over 40,000 miles—longer than Earth's circumference! It wraps around the globe and is the longest continuous mountain system on Earth.

FASTEST HUMAN-MADE OBJECT: NASA'S PARKER SOLAR PROBE

Launched in 2018, the Parker Solar Probe is the fastest human-made object, reaching speeds of 430,000 mph as it slingshots around the sun. That's fast enough to travel from New York to Tokyo in under a minute.

UNUSUAL WORLD RECORDS IN FOOD, FASHION, AND FUN

WORLD'S LARGEST PIZZA WAS BIGGER THAN A BASKETBALL COURT

In 2023, YouTuber Airrack and Pizza Hut created the world's largest pizza in Los Angeles, measuring a whopping 13,990 square feet. It took over 13,000 pounds of dough, and it was baked in pieces with heat guns.

LONGEST FINGERNAILS ON A PAIR OF HANDS EVER

American Lee Redmond grew her nails for nearly 30 years. Her nails reached a combined length of 28 feet, 4.5 inches before she lost them in a car accident. She carefully manicured and painted them to maintain the record.

WORLD'S LARGEST COLLECTION OF RUBBER DUCKS

Charlotte Lee from Washington has the world's largest collection of rubber ducks, totaling over 9,000 distinct ducks. She began collecting in 1996 and has an entire room dedicated to her flock.

MOST SPOONS BALANCED ON A HUMAN BODY

Etibar Elchiyev of Georgia balanced 79 metal spoons on his body in 2013. He trained his posture and steadiness for months to earn this strange but impressive record.

RECORDS SET BY MISTAKE OR FLUKE

THE LONGEST TENNIS MATCH WASN'T SUPPOSED TO HAPPEN

In 2010, John Isner and Nicolas Mahut played the longest tennis match in history at Wimbledon. It lasted 11 hours and 5 minutes. Neither player expected the match to go beyond a few hours, but it ended up stretching over three days. Their serves were so strong and consistent that neither could break the other's game for 183 consecutive games.

THE FIRST PHOTOGRAPH OF A HUMAN WAS A FLUKE

The earliest known photograph of a person wasn't taken intentionally. In an 1838 Paris street photo by Louis Daguerre, a man getting his boots shined stood still long enough to be captured in the long exposure, accidentally becoming the first photographed human in history.

TYPO LED TO THE LONGEST NAME OF A PLACE

Taumatawhakatangihangakoauauotamateaturipukakapikimaung ahoronukupokaiwhenuakitanatahu (85 letters) is the name of a hill in New Zealand. However, part of its length was due to early mapmakers including extra Māori descriptors, some possibly duplicated or exaggerated by accident.

THE LOUDEST SOUND EVER RECORDED WAS AN ACCIDENT

In 1883, the eruption of Krakatoa produced the loudest sound in recorded history. It was heard over 3,000 miles away. Ships' crews and people in distant cities heard what they thought were cannon blasts. In reality, it was a catastrophic volcanic eruption that changed the global climate for years.

A WRONG TURN SET THE FIRST AROUND-THE-WORLD FLIGHT RECORD

In 1924, the first flight around the world by U.S. Army pilots almost didn't happen. A wrong turn early in the journey nearly caused one of the planes to crash, and several had to be replaced. Despite mechanical failures and accidents, the team completed the 175-day trip and set a record—for resilience as much as distance.

STRANGE GUINNESS RECORDS NO ONE ASKED FOR

LONGEST DISTANCE TO PULL AN AIRPLANE WITH A BEARD

In 2013, Antanas Kontrimas of Lithuania pulled an 8,000-pound aircraft over 11 feet using just his beard. Not recommended by hair-care professionals or chiropractors.

MOST SNAILS ON A FACE

In 2007, Fin Kehler, a 10-year-old from Utah, had 43 snails placed on his face for 10 seconds to break the record. His only challenge? Not flinching while they slimed around.

FASTEST TIME TO EAT A BOWL OF PASTA WITH NO HANDS

In 2017, Michelle Lesco of Arizona devoured a bowl of pasta in 26.69 seconds without using her hands. It's unclear whether marinara stains were part of the judging criteria.

MOST RATTLESNAKES HELD IN THE MOUTH

Jackie Bibby holds the record for holding 13 live rattlesnakes by their tails in his mouth. He sat calmly with them dangling for over 10 seconds. He's also known as "The Texas Snake Man," which checks out.

LARGEST ANIMALS, STRUCTURES, AND COLLECTIONS

THE LARGEST ANIMAL EVER: THE BLUE WHALE

The blue whale is the largest animal to have ever lived, bigger than any dinosaur. It can reach 100 feet in length and weigh up to 200 tons. It has a heart the size of a small car and blood vessels wide enough for a human to crawl through.

THE LARGEST LIVING STRUCTURE: THE GREAT BARRIER REEF

Stretching over 1,400 miles, the Great Barrier Reef is the largest living structure on Earth. It's even visible from space. Technically, it's a massive colony of tiny coral polyps that's home to over 9,000 species of marine life.

THE LARGEST SPIDER WEB EVER RECORDED

In 2009, researchers in Madagascar discovered a Darwin's bark spider web that spanned 82 feet—longer than a school bus—and crossed a river. By weight, the spider's silk is tougher than Kevlar.

THE LARGEST KNOWN FUNGUS: THE HUMONGOUS FUNGUS

In Oregon, an underground network of *Armillaria ostoyae* mushrooms spans over 3.4 square miles and is estimated to be over 2,400 years old. Although mostly hidden underground, it's technically the largest living organism by area.

THE WORLD'S LARGEST COLLECTION OF TAXIDERMY FROGS DRESSED AS HUMANS

In Croatia, the Froggyland Museum houses 507 real frogs posed in scenes such as schools, circuses, and sewing shops. All frogs are over a century old and were preserved by a Hungarian taxidermist with very specific interests.

MOST EXPENSIVE OR VALUABLE ITEMS IN THE WORLD

THE MOST EXPENSIVE PAINTING EVER SOLD

Leonardo da Vinci's *Salvator Mundi* was sold for a staggering $450.3 million in 2017. It was believed lost for centuries and had even been stored in a Louisiana kitchen before being restored and auctioned to a Saudi prince.

THE WORLD'S MOST VALUABLE COIN

The 1933 Double Eagle $20 gold coin is worth over $18.8 million. Although 445,500 were minted, they were never circulated due to FDR's gold recall. Nearly all were melted down, and owning one was illegal — until one slipped through.

THE MOST EXPENSIVE SUBSTANCE

At around $62.5 trillion per gram, antimatter is the most expensive substance on Earth. Produced in tiny quantities in particle accelerators, it could theoretically become a future fuel for space travel if we ever figure out how to contain it.

THE MOST EXPENSIVE CAR EVER SOLD AT AUCTION

A 1955 Mercedes-Benz 300 SLR Uhlenhaut Coupé sold for $143 million in 2022, making it the most valuable car ever auctioned. Only two were ever made, and one had been sitting in the company's collection for decades.

THE MOST VALUABLE BASEBALL

The home run ball hit by Mark McGwire for his 70th home run in the 1998 season sold for $3 million in 1999. After steroid scandals, its value plummeted. Today, it's worth just a fraction, but it's still a powerful symbol of sports history and hype.

FIRSTS IN HUMAN ACHIEVEMENT ACROSS DISCIPLINES

FIRST WOMAN TO WIN A NOBEL PRIZE

In 1903, Marie Curie became the first woman to win a Nobel Prize and later became the first person to win two (in physics and chemistry). Her pioneering work on radioactivity laid the foundation for modern physics and cancer treatments.

FIRST MAN TO CONQUER EVEREST

On May 29, 1953, Sir Edmund Hillary of New Zealand and Tenzing Norgay, a Sherpa of Nepal, were the first climbers confirmed to reach the summit of Mount Everest, the world's tallest peak. They kept it humble, each crediting the other.

FIRST POWERED FLIGHT

On December 17, 1903, Orville and Wilbur Wright achieved the first powered, controlled flight in Kitty Hawk, North Carolina. The flight lasted 12 seconds and traveled 120 feet—shorter than a Boeing 747's wingspan but revolutionary nonetheless.

FIRST COMPUTER PROGRAMMER

In the 1840s, long before computers existed, Ada Lovelace wrote the first algorithm intended for a machine, making her the world's first computer programmer. She imagined computing not just as math but as music and art too.

CHAPTER SIX:
SPORTS

THE ORIGINS AND EVOLUTION OF MAJOR SPORTS

SOCCER DATES BACK OVER 2,000 YEARS

A game called cuju, played in ancient China as early as the second century BCE, involved kicking a leather ball through a small opening. It's considered the earliest known form of soccer. Other versions also existed in Mesoamerica, Japan, and Greece.

BASKETBALL WAS INVENTED TO CURE BOREDOM

In 1891, Canadian teacher James Naismith invented basketball using a soccer ball and two peach baskets. He needed an indoor activity to keep his students busy during snowy Massachusetts winters, and the rest is hoops history.

THE FIRST OLYMPICS WERE WILDLY DIFFERENT

The original Olympic Games in ancient Greece took place in 776 BCE and had just one event: a footrace. Athletes competed naked, and the games were held to honor Zeus. Over time, wrestling, javelin, and even chariot races were added.

BASEBALL'S RULES WERE ONCE A FREE-FOR-ALL

Early forms of baseball in 19th-century America had no standard rules. Some teams counted "soaking" (hitting the runner with the ball) as an out. The Knickerbocker Rules, written in 1845, helped formalize the game we know today.

AMERICAN FOOTBALL EVOLVED FROM RUGBY AND SOCCER

American football took its own form thanks to innovations like the forward pass and structured plays. It was influenced heavily by figures such as Walter Camp, the "Father of American Football."

OBSCURE OR BANNED SPORTS THROUGHOUT HISTORY

FOX TOSSING (YES, REALLY)

In 17th- and 18th-century Europe, aristocrats played a gruesome sport where couples used slings to launch live animals—often foxes—into the air. The "winner" was whoever flung the animal highest. It was eventually banned due to its cruelty.

SHIN KICKING WAS ONCE AN ENGLISH PASTIME

An actual event in England's "Cotswold Olimpicks" since the early 1600s, shin kicking involves two competitors stuffing their pants with straw and trying to kick each other in the shins until one of them falls. It still exists today but with a few safety rules.

AFGHANISTAN'S BRUTAL NATIONAL SPORT

Still played today, buzkashi is a Central Asian sport in which horseback riders compete to drag a headless goat carcass across a field. Once banned by the Taliban, it's a deeply rooted tradition that symbolizes strength, strategy, and honor.

OCTOPUS WRESTLING WAS ONCE A REAL THING

In the 1950s and 1960s, octopus-wrestling competitions were held in the Pacific Northwest. Divers would grapple with giant Pacific octopuses and drag them to the surface. It faded as marine conservation awareness grew.

LA SOULE: MEDIEVAL CHAOS ON A FIELD

A forerunner to rugby and soccer, La Soule was a violent, loosely organized ball game played in France and the British Isles. Townsfolk clashed with few rules—sometimes with hundreds of players—and the ball could be a pig bladder, a wooden orb, or even a rock.

ATHLETES WITH STRANGE PRE-GAME RITUALS

RAFAEL NADAL'S WATER BOTTLE

Tennis legend Rafael Nadal is famous for meticulously lining up his water bottles parallel to the court, labels facing outward, before high-pressure matches. Nadal expresses that the ritual helps him feel in control and grounded.

SERENA WILLIAMS BOUNCES HER WAY TO VICTORY

Before every match, Serena Williams follows a strict routine: She brings her shower sandals to the court, ties her shoelaces in a specific way, and bounces the ball exactly five times before her first serve. She says breaking her rituals throws off her game.

WADE BOGGS: CHICKEN EVERY DAY

Baseball Hall-of-Famer Wade Boggs was known as the "Chicken Man" because he ate chicken before every game. He also ran sprints at 7:17 p.m., wrote the word *chai* in the dirt, and took batting practice at precisely 5:17 p.m. every game day.

MICHAEL JORDAN'S LUCKY SHORTS

Throughout his NBA career, Michael Jordan wore his University of North Carolina shorts under his Chicago Bulls uniform for good luck. The superstition lasted for years and even influenced NBA uniform styles, leading to longer, baggier shorts.

MOISES ALOU SKIPPED THE GLOVES AND USED URINE

MLB outfielder Moises Alou refused to wear batting gloves. Instead, he urinated on his hands before games, claiming it toughened his skin and improved grip.

OLYMPIC ODDITIES, CONTROVERSIES, AND FORGOTTEN EVENTS

THE MARATHONER WHO TOOK A CAR RIDE

During the 1904 Olympic marathon, runner Fred Lorz dropped out due to exhaustion, then hitched a ride to the finish line. Intended as a joke, Lorz jogged across the finish line and was temporarily declared the winner before being exposed.

ART USED TO BE AN OLYMPIC EVENT

From 1912 to 1948, the Olympics awarded medals for painting, sculpture, music, architecture, and literature. These "artistic competitions" were removed because many artists were considered professionals, which contradicted Olympic ideals, as the rules require participants to be amateurs.

THE "NAKED" SKI JUMPER

Michael Edwards became an endearing symbol for the underdog after finishing dead last in the 1988 Olympics. Edwards's ambitious spirit and lack of proper gear rallied spectators and went on to inspire the 2015 film *Eddie the Eagle*.

THE MISSING GOLD MEDALIST

In the 1900 Games, Belgian rower François Brandt and his teammate replaced their too-heavy coxswain with a local boy. The boy helped them win gold but disappeared into history. His identity was never recorded, and he remains the youngest unknown Olympic champion.

THE 1936 NAZI SALUTE MIX-UP

During the 1936 Berlin Olympics, American sprinter Jesse Owens famously beat Nazi Germany's athletes. Some photos show him raising his hand, leading to a myth that he gave a Nazi salute, but he was actually waving to the crowd. Owens won four gold medals, defying Hitler's racist ideology.

STRANGE GAME RULES YOU DIDN'T KNOW EXISTED

YOU CAN WIN WITHOUT EVER MOVING

In chess, if your opponent forfeits or breaks the rules before you make a move, you can technically win without touching a single piece.

SOCCER GOALKEEPERS' SIX-SECOND RULE

In professional soccer, a goalie is technically only allowed to hold the ball for six seconds. Break it, and it should result in an indirect free kick, but referees almost never call it, making it one of the strangest "invisible" rules in sports.

THE SERVE MUST BE TOSSED EXACTLY SIX INCHES

In official table tennis matches, the ball must be thrown at least six inches straight up during a serve. Failing to do so can result in a fault. Also, you can't hide the ball with your hand during the toss.

YOU CAN STEAL FIRST BASE

In some minor league games experimenting with new rules, batters can steal first base on any wild pitch or passed ball, even if it's not strike three. It's a rare occurrence, but it's totally legal under those trial rules.

A MISS CAN RESET THE TABLE

In snooker (a form of billiards), if a player fails to hit the intended ball, referees can declare a "miss" and the opponent can force the shot to be replayed. In theory, this can go on forever if the referee keeps calling fouls and players keep missing.

FAMOUS SPORTS UPSETS AND UNDERDOG STORIES

MIRACLE ON ICE AS USA BEATS SOVIET UNION

The U.S. men's hockey team, made up of college amateurs, defeated the heavily favored Soviet team, which had dominated international play for years. The 4–3 win in Lake Placid became one of the most iconic moments in sports history, and it wasn't even the final game of the tournament!

LEICESTER CITY WINS THE PREMIER LEAGUE AT 5,000-TO-1 ODDS

English soccer club Leicester City, expected to be relegated, shocked the world by winning the Premier League title. Bookmakers had given them 5,000-to-1 odds, the same as betting on Elvis being found alive.

BUSTER DOUGLAS KNOCKS OUT MIKE TYSON

Boxing legend Mike Tyson was undefeated and feared around the world until James "Buster" Douglas, a 42-to-1 underdog, knocked him out in the 10th round. It remains one of the greatest upsets in boxing history.

RULON GARDNER BEATS THE UNBEATABLE

In the Sydney Games, American wrestler Rulon Gardner defeated Aleksandr Karelin, a Russian who'd been undefeated for 13 years and hadn't given up a point in six. Gardner pulled off one of the most shocking gold medal wins in Olympic history.

UNBELIEVABLE RECORDS IN SPORTS

MOST GOALS IN A SINGLE SOCCER MATCH

In 2002, Madagascar's team AS Adema won a match 149–0 against Stade Olympique de l'Emyrne, but it wasn't a true blowout. The losing team intentionally scored their own goals in protest of a refereeing decision. Still, it stands as the highest-scoring soccer game in history.

MOST OLYMPIC MEDALS

Swimmer Michael Phelps holds the record for the most Olympic medals of all time: 23 gold, 3 silver, and 2 bronze. No other athlete comes close, making him arguably the greatest Olympian in history.

FASTEST 100-METER SPRINT ON ALL FOURS

Yes, this is real. Japanese athlete Kenichi Ito holds the record for the fastest 100 meters on all fours, completing it in 15.71 seconds. He studied primates for years to perfect his running technique!

LONGEST WINNING STREAK IN PRO SPORTS

Pakistani squash legend Jahangir Khan won 555 matches in a row between 1981 and 1986. That's the longest winning streak in professional sports history, a nearly unbeatable record in terms of dominance.

HOW TECHNOLOGY IS CHANGING MODERN SPORTS

WEARABLE TECH TRACKS EVERYTHING

Athletes now use wearable sensors that monitor heart rate, hydration, lactic acid, and even sweat composition in real time. Some devices, like smart patches and vests, give coaches a live feed of how an athlete's body is performing and recovering during a game.

VIDEO ASSISTANT REFEREE AND INSTANT REPLAY ARE CHANGING THE RULES OF THE GAME

Video Assistant Referee technology has completely changed decision-making in soccer, football, and tennis. While controversial, it helps ensure critical calls like goals and fouls are as accurate as possible, sometimes overturning centuries-old traditions of trusting the referee's word alone.

ESPORTS AND AI TRAINING ARE CREATING NEW KINDS OF ATHLETES

Artificial intelligence is now being used to simulate opponents and analyze gameplay in esports, martial arts, and even chess. Players can train against AI that mimics top-tier athletes, allowing them to prepare more strategically than ever before.

SMART STADIUMS USE FACIAL RECOGNITION AND AUGMENTED REALITY

Modern sports arenas are becoming high-tech environments. Some stadiums use facial recognition for security, augmented reality for replays, and app-based concessions that let fans order food without leaving their seats.

3D PRINTING AND BIONIC GEAR ARE CUSTOMIZING THE ATHLETE

From 3D-printed running shoes to carbon-fiber exoskeletons for training, athletes now benefit from custom-designed gear that enhances speed and protection. Some disabled athletes are even outperforming able-bodied ones with the help of advanced prosthetics.

CHAPTER SEVEN: POP CULTURE

HIDDEN EASTER EGGS
IN MOVIES AND SHOWS

PIXAR'S A113 IS IN NEARLY EVERY FILM

The code A113 shows up in almost every Pixar and Disney animated movie, whether on license plates, doors, or signs. It's a nod to Room A113 at CalArts, where many Pixar animators studied.

R2-D2 AND C-3PO APPEAR IN *INDIANA JONES*

In *Raiders of the Lost Ark*, hieroglyphs on a wall briefly show R2-D2 and C-3PO. It's a cheeky nod from George Lucas and Steven Spielberg linking Star Wars and Indiana Jones.

FIGHT CLUB HAS A STARBUCKS CUP IN ALMOST EVERY SCENE

Director David Fincher placed a Starbucks cup in nearly every scene of *Fight Club* as a satire of corporate consumerism. Once you notice it, you can't unsee it.

BREAKING BAD'S CHEMISTRY

The episode titles in Breaking Bad season two secretly foreshadow a major event. When you combine the four episodes "Seven Thirty-Seven," "Down," "Over," and "ABQ," it hints at the plane crash that ends the season.

THE PIZZA PLANET TRUCK IS EVERYWHERE

Pixar's Pizza Planet delivery truck makes at least one appearance in every Pixar movie, even in historical or fantastical settings. In *Brave* (set in medieval Scotland), it's carved into a piece of wood!

MUSIC FACTS ABOUT FAMOUS SONGS AND ARTISTS

"HEY JUDE" WAS WRITTEN FOR JOHN LENNON'S SON

Paul McCartney wrote "Hey Jude" to comfort Julian Lennon during his parents' divorce. Originally titled "Hey Jules," it was meant to reassure him during the difficult time but turned into one of the Beatles' biggest hits.

PRINCE PLAYED 27 INSTRUMENTS ON HIS DEBUT ALBUM

At just 20 years old, Prince released *For You* and played every single instrument—27 in total—on the album. That included guitar, drums, bass, keyboards, and even a gong.

"THE SOUND OF SILENCE" WAS ACCIDENTALLY REMIXED INTO A HIT

Simon & Garfunkel's "The Sound of Silence" originally flopped until a producer added electric instruments and drums without their permission. The remix topped the charts and launched their career.

NIRVANA'S "SMELLS LIKE TEEN SPIRIT" TITLE CAME FROM A JOKE

The title came from graffiti a friend wrote on Kurt Cobain's wall: *Kurt smells like Teen Spirit*. Cobain didn't know it was a reference to a brand of deodorant; he thought it sounded revolutionary.

BEYONCÉ RECORDED "HALO" IN THREE HOURS

One of Beyoncé's most iconic ballads, "Halo," was recorded in under three hours. She nailed it so quickly that producer Ryan Tedder called it "one-take magic."

UNEXPECTED CROSSOVERS AND CAMEOS IN MEDIA

SPIDER-MAN ALMOST JOINED THE X-MEN

Before superhero studios were divided, there were plans for Spider-Man to appear in *X-Men* (2000). In fact, Hugh Jackman showed up on the *Spider-Man* set hoping to shoot a Wolverine cameo, but it was scrapped due to costume issues.

STEPHEN KING'S BOOKS ARE ALL CONNECTED

Many of Stephen King's novels—like *It*, *The Shining*, and *The Dark Tower*—exist in a shared multiverse with characters, places, and cosmic forces bleeding into each other. Even Pennywise and Roland the Gunslinger are linked through this sprawling mythology.

R2-D2 AND E.T. EXIST IN THE SAME UNIVERSE

In *Star Wars: The Phantom Menace*, you can spot E.T.'s alien species in the Galactic Senate. And in *E.T.*, a kid wears a Yoda costume, prompting E.T. to recognize him. Spielberg and Lucas were trading Easter eggs, suggesting a shared universe across galaxies!

MICHAEL JACKSON WANTED TO BE SPIDER-MAN—LITERALLY

In the 1990s, Michael Jackson tried to buy Marvel Comics so that he could cast himself as Spider-Man in a film adaptation. He was a huge comic fan and envisioned himself swinging across the screen.

TONY STARK'S AI WAS VOICED BY PAUL BETTANY

Before becoming Vision in the *Avengers* films, Paul Bettany voiced J.A.R.V.I.S., Tony Stark's AI assistant. The character evolved into Vision later in the series, making Bettany one of the few actors to play two major roles within the same character arc.

WILD ORIGIN STORIES OF FAMOUS CHARACTERS

SHREK WAS ORIGINALLY MEANT FOR CHRIS FARLEY

Before Mike Myers took the role, Chris Farley recorded nearly all the dialogue for Shrek. His version was more soft-spoken and emotional. But after his tragic death, the script was rewritten, and Myers gave the ogre his now-iconic Scottish accent.

DARTH VADER WASN'T SUPPOSED TO BE LUKE'S FATHER

In George Lucas's original drafts of Star Wars, Darth Vader and Anakin Skywalker were two separate characters. The famous twist — "I am your father" — was added in later drafts, changing the saga's emotional core forever.

BATMAN NEARLY HAD A RED SUIT AND WINGS

In his earliest sketches, Batman had a red costume, stiff wings, and no cowl. Creator Bob Kane's idea was radically reworked by writer Bill Finger, who gave him the bat motif, cape, and darker tone we now know.

THE JOKER WAS INSPIRED BY A SILENT FILM AND A PLAYING CARD

The Joker's look was heavily influenced by the 1928 silent film *The Man Who Laughs*, starring Conrad Veidt as a man with a permanent grin carved into his face. Add a deck of playing cards and the Clown Prince of Crime was born.

WOLVERINE WAS ORIGINALLY A TEENAGE MUTANT

When first conceived, Wolverine was going to be a mutated wolverine cub evolved into a human by the High Evolutionary. Thankfully, Marvel scrapped the animal origin and turned him into a mysterious, brooding mutant with claws and a dark past.

POP CULTURE MYTHS
DEBUNKED

WALT DISNEY WAS *NOT* CRYOGENICALLY FROZEN

A long-standing urban legend claims that Walt Disney had his body frozen after death to be revived in the future. In truth, he was cremated, and there's no evidence he ever expressed interest in cryonics.

"BEAM ME UP, SCOTTY" WAS NEVER ACTUALLY SAID IN *STAR TREK*

Despite being one of the most quoted sci-fi lines ever, no character ever says "Beam me up, Scotty" in the original Star Trek series. Closest versions were "Scotty, beam me up" or "Beam us up, Mr. Scott."

MARILYN MANSON WASN'T IN *THE WONDER YEARS*

The myth that Marilyn Manson played Paul Pfeiffer on *The Wonder Years* is completely false. Paul was actually played by actor Josh Saviano, who later became a lawyer, not a goth icon.

THE MCDONALD'S MONOPOLY GAME WASN'T JUST RIGGED BY LUCK; IT WAS AN INSIDE JOB

Many believed the *Monopoly* sweepstakes was impossible to win just by chance, and they were right. For over a decade, an inside man stole winning pieces and gave them to accomplices, leading to an FBI investigation and arrests in 2001.

THE RED PILL IN *THE MATRIX* ISN'T ABOUT POLITICS

Many online groups have co-opted the term *red pill* to represent a political awakening, but this was never the original intent. In fact, *The Matrix* creators, the Wachowskis, later revealed that the red pill was a metaphor for gender identity and personal transformation, reflecting their own journeys as trans women. The modern reinterpretation is far removed from its cinematic and philosophical roots.

BEHIND-THE-SCENES FILM FACTS AND BLOOPERS

THE CHEST-BURSTER SCENE IN *ALIEN* WAS A COMPLETE SURPRISE

To capture genuine shock, director Ridley Scott didn't tell the cast exactly what would happen during the infamous chest-bursting scene. The actors' horrified reactions—especially Veronica Cartwright's scream—were 100% real.

VIGGO MORTENSEN BROKE HIS TOE IN *THE TWO TOWERS*

When Aragorn kicks a helmet and lets out a primal scream in *The Two Towers*, it wasn't just acting. Viggo Mortensen broke two toes doing it and stayed in character. That scream is the real deal.

DUSTIN HOFFMAN'S FAMOUS "I'M WALKIN' HERE!" WAS IMPROVISED

In *Midnight Cowboy*, when a cab nearly hit Dustin Hoffman and Jon Voight, Hoffman stayed in character and shouted, "I'm walkin' here!" The unscripted moment became one of the most iconic lines in film history.

STORMTROOPER ACCIDENTALLY HIT HIS HEAD IN STAR WARS

In *A New Hope*, one Stormtrooper famously bumps his head on a door. It was a real blooper, but George Lucas found it funny enough to leave it in the final cut and even added a sound effect in the remastered edition.

THE SLIPPERY FLOOR WASN'T PART OF THE SCENE IN *DIE HARD*

Bruce Willis's iconic scene where John McClane slides across a floor in Die Hard wasn't planned. The floor was accidentally wet, and Willis slipped mid-take. It looked cool, so they kept it.

VIRAL TRENDS AND WHERE THEY CAME FROM

THE HARLEM SHAKE WASN'T INVENTED BY THE INTERNET

The viral Harlem Shake dance trend exploded in 2013 thanks to YouTube, but the original Harlem Shake is a real dance that originated in 1981 in Harlem, New York. The internet version looked nothing like the real thing, leading to backlash from the original creators and community.

RICKROLLING STARTED AS A VIDEO GAME FORUM PRANK

Rickrolling involves tricking someone into clicking a link to Rick Astley's "Never Gonna Give You Up." It started in 2007 on 4chan, where users baited others with fake news about *Grand Theft Auto IV*. The bait-and-switch prank exploded from there and is still going strong.

PLANKING WAS INVENTED IN THE 1990S

The odd trend of lying stiff like a plank in random places seemed new in the 2010s, but it actually started in 1997 when two guys in England called it "The Lying Down Game." It didn't go viral until Facebook picked it up over a decade later.

THE ICE BUCKET CHALLENGE RAISED OVER $115 MILLION

What started as a viral dare on social media turned into one of the most successful charity campaigns ever. In 2014, the Ice Bucket Challenge raised over $115 million for ALS research, massively boosting awareness and directly funding breakthroughs in treatment.

TIKTOK DANCE TRENDS OFTEN COME FROM TEEN CREATORS WHO DON'T GET CREDIT

Many of the biggest TikTok dances—like the *Renegade*—were created by young, often Black, creators such as Jalaiah Harmon. However, they only gained traction when reposted by influencers who got all the attention. This led to a cultural reckoning around digital ownership and credit in viral trends.

72

FORGOTTEN FADS FROM PAST DECADES

POGS TOOK OVER THE PLAYGROUND

Little cardboard discs called pogs were based on a 1930s game from Hawaii that used caps from milk bottles. They exploded in popularity in the mid-1990s, even causing school bans due to gambling concerns before disappearing just as fast.

PET ROCKS MADE MILLIONS

In 1975, Gary Dahl marketed ordinary rocks as "pets," complete with boxes, breathing holes, and training manuals. It was a satirical commentary on consumer culture, and it worked. He sold over 1.5 million Pet Rocks in just a few months.

TOMAGOTCHIS WERE THE ORIGINAL DIGITAL TETHER

These pocket-sized virtual pets from Japan launched globally in the late 1990s. Kids had to feed, play with, and clean up after their Tamagotchis constantly or risk digital death. Many schools banned them for being too distracting and emotionally intense.

MOOD RINGS WERE THE 1970S VERSION OF A PERSONALITY QUIZ

Invented in 1975, mood rings were supposed to change color based on your emotional state. In reality, they just reacted to body temperature, but they became a huge hit as a symbol of self-expression during the "New Age" boom.

"MACARENA" WAS A GLOBAL PHENOMENON

Originally released in 1993 by the Spanish duo Los Del Río, the song didn't catch fire until a remix by the Bayside Boys hit the U.S. charts in 1996. It became a dance craze at weddings, schools, and sports events worldwide, then disappeared as quickly as it came.

CHAPTER EIGHT: WEIRD LAWS AND CUSTOMS

OUTDATED OR BIZARRE LAWS STILL IN ACTION

KENTUCKY: IT'S ILLEGAL TO CARRY AN ICE CREAM CONE IN YOUR BACK POCKET

This strange law was originally created to deter horse theft. Thieves would lure horses away using an ice cream cone placed in their back pocket, and since the horse followed rather than being led, it wasn't considered theft under older laws!

ARIZONA: IT'S ILLEGAL FOR A DONKEY TO SLEEP IN A BATHTUB

This law stems from a 1920s flooding incident when a rancher's donkey was sleeping in a bathtub and had to be rescued after a dam broke. It caused so much chaos that a law was passed to prevent future bathtub donkey rescues.

DENMARK: YOU MUST CHECK UNDER YOUR CAR FOR SLEEPING CHILDREN

While not strictly enforced today, an old Danish law required drivers to look beneath their car before starting it. The rule was likely tied to child safety from a time when kids often played in the streets or near parked cars.

FLORIDA: UNMARRIED WOMEN CAN'T SKYDIVE ON SUNDAYS

Although not enforced anymore, this bizarre and sexist law from Florida technically bans unmarried women from parachuting on

Sundays. It's possibly tied to early 20th-century notions of decency and gender roles.

UNITED KINGDOM: IT'S ILLEGAL TO PRETEND TO BE A WITCH BUT NOT TO BE ONE

Under the Witchcraft Act of 1735, pretending to have magical powers or tell fortunes for money was a criminal offense, while actually being a witch wasn't technically illegal. The law was used even in the 20th century, most notably in 1944 when a woman named Helen Duncan was jailed for "pretending to conjure spirits" during WWII. The law was finally repealed in 1951.

STRANGE BANS AROUND THE WORLD

CHEWING GUM IS (MOSTLY) BANNED IN SINGAPORE

Since 1992, Singapore has strictly banned the import and sale of chewing gum to keep public spaces clean and infrastructure like subway doors functional. Only therapeutic gum such as nicotine gum is allowed, and even that requires a prescription!

WINNIE THE POOH IS BANNED FROM POLISH PLAYGROUNDS

In certain towns in Poland, Winnie the Pooh has been banned from being used as a mascot in playgrounds due to concerns that he's "inappropriate" for children because he doesn't wear pants and is "gender ambiguous." Yes, really.

REINCARNATION WITHOUT STATE PERMISSION IS BANNED IN TIBET

The Chinese government passed a 2007 law stating that Buddhist monks in Tibet cannot reincarnate without government approval. The rule was intended to control the selection of spiritual leaders, especially the next Dalai Lama.

IT'S ILLEGAL TO OWN JUST ONE GUINEA PIG IN SWITZERLAND

In Switzerland, guinea pigs are considered social animals, so owning just one is deemed animal abuse. You must have at least two or you're breaking the law. They even have rent-a-guinea-pig services for when one dies!

BABY NAMES ARE HEAVILY REGULATED IN ICELAND

Iceland has a Naming Committee that must approve all baby names to ensure they fit Icelandic language and grammar rules. If a name isn't on the official list, parents must apply for special permission. *Harriet* and *Enrique* have both been denied!

CULTURAL TRADITIONS THAT SEEM ODD TO OUTSIDERS

AMPUTATING FINGERS WHILE GRIEVING

In the Dani tribe of Papua, Indonesia, it was once customary for women to cut off part of a finger when a loved one died. This painful ritual symbolized deep mourning and emotional suffering.

The practice has largely been phased out, but elders still bear the scars.

BABY TOSSING

In the Torajan culture, deceased family members are often kept in the home for months or even years after death. They're cared for, dressed, and spoken to as though still alive until the family can afford an elaborate funeral.

NIGHT HUNTING

Known as *bomena*, this traditional Bhutanese courtship practice involved young men sneaking into the homes of women at night to try and win their affection, often through whispered conversation or song. While increasingly rare, it was once widely accepted.

WEARING RED AT FUNERALS

In some South African communities, red is worn at funerals instead of black. It symbolizes the shedding of blood and mourning but also life and rebirth — a powerful dual meaning that may surprise outsiders expecting somber colors.

UNUSUAL FESTIVALS AND CELEBRATIONS

LA TOMATINA

Held annually in Buñol, Spain, La Tomatina is a massive tomato fight where participants hurl over 100 metric tons of overripe tomatoes at each other. The festival started in 1945 as a spontaneous food fight during a parade and has since grown into a world-famous event.

GOOSE PULLING

In this centuries-old tradition, riders on horseback attempt to pull the head off a suspended goose as they gallop past. The winner is crowned "king" for the year. Today, many towns use dead geese or substitutes, and the practice is fading.

KANAMARA MATSURI

This Shinto festival in Kawasaki celebrates fertility and sexual health, centered around large pink and steel phallus statues. The event includes parades, candy, and offerings. It's also used to raise awareness and funds for HIV research.

EL COLACHO

Known as the "baby-jumping festival," El Colacho has men dressed as devils leaping over rows of babies lying on mattresses. The tradition, dating back to 1620, is meant to ward off evil spirits and ensure the infants' safe future.

MONKEY BUFFET FESTIVAL

In Lopburi, Thailand, locals hold a grand feast for monkeys. Participants stack tables with fruits, vegetables, and desserts to honor the macaques that roam the town. It's seen as an offering of goodwill, and it also draws a lot of tourism.

GOVERNMENT RULES THAT ACCIDENTALLY CAUSED CHAOS

UNITED STATES: PROHIBITION (1920–1933)

The U.S. banned alcohol, hoping to reduce crime and improve morality. Instead, crime rates soared, bootlegging flourished, and powerful organized crime syndicates like Al Capone's empire thrived. It also cost the government billions in lost tax revenue.

AUSTRALIA: THE GREAT EMU WAR (1932)

In response to emus destroying crops in Western Australia, the government sent soldiers armed with machine guns to cull the birds. The emus easily outran the soldiers and the mission failed. It became a national joke and global curiosity.

CHINA: MAO'S FOUR PESTS CAMPAIGN (1958)

Mao Zedong declared war on sparrows, rats, flies, and mosquitoes—especially sparrows, which were blamed for eating grain. Citizens killed millions of sparrows. The result? Locust populations exploded, causing devastating crop failures and contributing to a famine that killed millions.

THE WEST BANK AND ISRAEL: DAYLIGHT SAVING DISASTER (1999)

During a period of tension, Palestinians and Israelis were using different clocks due to a disagreement on when to change to daylight saving time. A bomb was timed to go off at a certain hour, but the time discrepancy caused it to detonate early, killing the bombers instead of their intended targets.

ENGLAND: WINDOW TAX (1696–1851)

To raise funds, the English government taxed people based on how many windows their house had. The result? People bricked up their windows to avoid the tax, which contributed to poor ventilation and health problems, especially during disease outbreaks.

ODD ETIQUETTE RULES FROM DIFFERENT COUNTRIES

DON'T TIP IN JAPAN

In Japan, tipping is considered rude or confusing. Excellent service is expected as part of the job, and leaving extra money can actually offend or fluster the staff. Instead, gratitude is shown through polite language and respectful behavior.

DON'T SHOW THE SOLES OF YOUR FEET

In many cultures, especially in Thailand, India, and Islamic countries, showing the bottom of your feet or pointing them at someone is seen as extremely disrespectful. Feet are considered the dirtiest part of the body.

USE BOTH HANDS TO GIVE OR RECEIVE

When giving or receiving anything in South Korea (especially gifts or business cards), it's polite to use both hands. Using only one hand can seem dismissive or disrespectful.

DON'T CLINK GLASSES IN HUNGARY

In Hungary, clinking beer glasses is considered bad luck and even offensive. This stems from a legend that Austrian generals celebrated the execution of Hungarian revolutionaries in 1849 by clinking their beer mugs. Hungarians vowed not to do the same for 150 years, and even today, many still avoid it out of respect.

BURPING IS A FORM OF PRAISE

In parts of the Middle East, India, and rural China, a loud burp after a meal can be considered a compliment to the chef, showing that you are full and satisfied. But since it's not universally accepted, always check the local norms.

REAL LEGAL CASES
THAT MAKE NO SENSE

THE MAN WHO SUED HIMSELF

In 1995, a prisoner named Robert Lee Brock in Virginia sued himself for $5 million, claiming he violated his own civil rights by getting drunk and committing a crime. Since he was in jail and couldn't earn an income, he wanted the state to pay on his behalf. The judge dismissed the case but noted it was "an innovative approach."

THE FLYING SAUCER DIVORCE

In 2000, a Romanian man filed for divorce because his wife was abducted by aliens. She denied the event ever happened. The court ultimately granted the divorce because the couple's relationship had clearly broken down.

TOO HOT TO HANDLE

Often mocked, the 1992 *Liebeck v. McDonald's* case actually had serious merit. A woman suffered third-degree burns when she spilled 190°F coffee in her lap. She initially asked McDonald's to pay her medical bills; when they refused, she sued — and won. The case led to major changes in product warning labels, though it's widely misunderstood.

THE HAUNTED HOUSE LAWSUIT

In 1991, a buyer sued after learning the house he purchased in New York was advertised as haunted in the media, without being told by the seller. The court ruled in favor of the buyer, stating that since the seller had "ghost promoted" the home, they were legally obligated to disclose the haunting.

FALSE ADVERTISEMENT: BUDWEISER ISN'T A CHICK MAGNET

A man once sued Budweiser for false advertising, claiming their commercials gave the impression that drinking Bud Light would result in him being surrounded by beautiful women. He alleged mental and emotional distress when this didn't happen. Unsurprisingly, the court dismissed the case.

CHAPTER NINE:
TECHNOLOGY

INVENTIONS THAT CHANGED THE WORLD

THE WHEEL WASN'T ORIGINALLY USED FOR TRANSPORTATION

The oldest known wheel dates back over 5,000 years, but it wasn't invented for carts or chariots. It was used as part of a potter's wheel for shaping clay. It took centuries before wheels were adapted for moving people and goods.

THE INTERNET WAS ORIGINALLY A MILITARY PROJECT

The internet began in the late 1960s as ARPANET, a U.S. Defense Department program to create a communication system that could survive a nuclear attack. What started as a way for scientists to share data became the backbone of modern global life.

THE PRINTING PRESS SPARKED A REVOLUTION

Invented by Johannes Gutenberg around 1440, the printing press made books affordable and accessible. It helped spread knowledge during the Renaissance and fueled the Protestant Reformation, making it one of the most powerful inventions in human history.

THE LIGHT BULB WASN'T JUST EDISON'S IDEA

While Thomas Edison gets most of the credit, more than 20 inventors worked on electric light bulbs before him. Edison's genius was creating a long-lasting filament and a practical system for home lighting. His 1879 design made electric lighting a commercial reality.

THE EVOLUTION OF EVERYDAY TECH

THE FIRST CELL PHONE WEIGHED TWO AND A HALF POUNDS

In 1973, Martin Cooper made the first mobile phone call on a device that weighed two and a half pounds. It took 10 hours to charge for only 30 minutes of talk time. Today's smartphones are not only lighter but also have more computing power than NASA during the moon landing.

THE FIRST COMPUTER MOUSE WAS MADE OF WOOD

Invented by Douglas Engelbart in 1964, the original computer mouse was a small wooden box with two wheels. It completely changed the way we interact with computers and inspired future graphical interfaces.

TV REMOTES WERE ONCE WIRED

In the 1950s, early remote controls were connected to TVs via long cables. The first wireless remote, developed by Zenith, used light beams and was nicknamed "Flashmatic." Infrared tech came much later.

ANCIENT CIVILIZATIONS USED CANDLES AND WATER TO KEEP TIME

Before mechanical and digital clocks, ancient civilizations used dripping water or marked candles to track time. These methods evolved into pendulums, quartz movements, and now atomic clocks that are accurate to a billionth of a second.

TECH FLOPS THAT WERE AHEAD OF THEIR TIME

APPLE NEWTON (1993)

The Apple Newton was one of the first personal digital assistants. It could take handwritten notes, store contacts, and send faxes. Despite its revolutionary concept, it flopped due to high cost and poor handwriting recognition, but it paved the way for the iPhone and iPad years later.

GOOGLE GLASS (2013)

Google Glass introduced wearable augmented reality tech, voice commands, and a built-in camera. However, the public wasn't ready for it. Privacy concerns and a clunky design led to its quick retreat. But today, similar tech is thriving in healthcare and military applications.

WEBTV (1996)

WebTV allowed users to browse the internet and check email from their TV way before smart TVs and streaming boxes became common. It was slow, limited, and overshadowed by computers, but it anticipated modern smart home entertainment.

SONY BETAMAX (1975)

Betamax offered superior video quality to VHS but lost the format war due to shorter recording time and licensing missteps. Ironically, the core technology lived on in broadcast studios and security systems for decades.

MICROSOFT TABLET PC (2001)

Microsoft launched a pen-driven tablet PC long before the iPad, promoting it as the future of mobile computing. While it failed commercially, it introduced touchscreen computing and mobile productivity concepts that are now standard in tablets and two-in-one laptops.

WEIRD USES OF MODERN TECHNOLOGY

SMART TOILETS THAT ANALYZE YOUR HEALTH

In Japan and parts of South Korea, high-tech toilets can do much more than flush. Some models analyze your urine and stool for health markers like glucose levels or infections, potentially catching diseases before symptoms even appear.

3D-PRINTED MEAT AND ORGANS

Scientists are now using 3D printers to create lab-grown meat and even prototype human organs. While it sounds like sci-fi, this tech could one day solve food shortages and the organ donor crisis.

VIRTUAL THERAPY FOR PHOBIAS AND PTSD

Virtual reality (VR) isn't just for games. Therapists use VR to simulate anxiety-inducing scenarios like flying or public speaking to help patients gradually overcome their fears. It's also used in treating veterans with post-traumatic stress disorder.

AI-GENERATED RELIGIOUS SERMONS AND ART

Artificial intelligence has been used to write religious sermons, compose hymns, and even paint divinely inspired artworks. Some temples and churches have even experimented with AI-powered robot "priests."

DRONE BEEKEEPING AND POLLINATION

With bee populations declining, researchers have developed tiny pollination drones that mimic bees. These drones are coated in a sticky substance to collect and deposit pollen, offering a futuristic backup plan to help crops grow.

INTERNET MILESTONES AND VIRAL FIRSTS

THE FIRST VIRAL VIDEO WAS A DANCING BABY

In 1996, "Baby Cha-Cha-Cha" became the first viral video shared over email chains and websites. It even made appearances on Ally McBeal, proving early on that weird + cute = viral.

"ME AT THE ZOO" WAS YOUTUBE'S FIRST VIDEO

The very first YouTube video was uploaded by cofounder Jawed Karim in 2005. It featured him at the San Diego Zoo talking about elephants. The 18-second clip helped launch what is now the world's largest video platform.

THE FIRST TWEET EVER SENT

Jack Dorsey, one of Twitter's founders, posted "just setting up my twttr" in 2006, unknowingly launching what would become a major real-time news, meme, and political battleground platform. That tweet later sold as an NFT for over $2.9 million.

THE FIRST HASHTAG WAS USED IN 2007

Open-source advocate Chris Messina proposed using the pound symbol to group topics. His tweet read: "How do you feel about using # (pound) for groups. As in #barcamp [msg]?" People initially thought it was weird and ignored it.

"GANGNAM STYLE" WAS THE FIRST YOUTUBE VIDEO TO HIT ONE BILLION VIEWS

Psy's infectious K-pop anthem literally broke the platform's view counter. "Gangnam Style" maxed out the 32-bit integer used by YouTube at the time. Google had to upgrade YouTube's code to accommodate its global takeover.

STRANGE PATENTS AND INVENTIONS THAT ACTUALLY EXIST

THE BIRD DIAPER

Tired of cleaning up after your flying friend? A patented avian diaper allows birds to roam freely indoors without leaving droppings everywhere.

THE ANTI-EATING FACE MASK

This device, patented by a dentist, is a wire cage that fits around your mouth and locks to the back of your head. It was intended to help people stop snacking between meals. Spoiler: It didn't take off.

A MOTORIZED ICE CREAM CONE THAT SPINS

Why twist your wrist when eating soft serve? This battery-powered cone rotates the ice cream for you. It's a marvel of over-engineering that's ideal for the truly ice cream obsessed or very lazy.

HIGH-FIVE MACHINE LETS YOU FEEL LESS ALONE

Inventor Courtney Latham designed a machine that delivers a high five at the press of a button. Originally created for a class project, it became a viral symbol of silly yet strangely uplifting innovations.

THE BEARD BIB

A bib that suctions to your bathroom mirror to catch beard trimmings sounds ridiculous, but it's real and wildly popular on novelty sites. It's also saved many relationships from sink-hair rage.

FUTURISTIC IDEAS THAT ALREADY CAME TRUE

VIDEO CALLS WERE PREDICTED IN 1911

In *Tom Swift and His Photo Telephone* (1911), author Victor Appleton describes a device that lets people see and hear each other from afar. A century later, FaceTime, Zoom, and Skype made that idea a basic part of modern communication.

TABLETS AND E-READERS FIRST APPEARED IN 1968

Stanley Kubrick's film *2001: A Space Odyssey* showed astronauts using touchscreen devices for reading and entertainment decades before the first iPad launched in 2010. Apple even referenced the film in court during a lawsuit about tablet design!

JETPACKS ACTUALLY EXIST

Jetpacks have been a staple of sci-fi for years, and while we're not all commuting through the skies, working jetpacks do exist. Companies like JetPack Aviation and Gravity Industries are developing them for military, rescue, and recreational use.

VOICE ASSISTANTS LIKE SIRI AND ALEXA WERE IMAGINED IN THE 1960S

Star Trek's crew spoke to their computer as if it were a helpful assistant. Now, we casually ask Siri for weather updates or tell Alexa to play music, bringing that futuristic concept right into our homes.

BIONIC LIMBS RESPOND TO BRAIN SIGNALS

Modern prosthetics can now be controlled by a user's brainwaves or muscle signals. These bionic limbs allow users to grasp, move, and feel, restoring function in ways once thought impossible.

CHAPTER TEN:
SPACE

MYSTERIOUS MOONS, PLANETS, AND GALAXIES

EUROPA MIGHT HARBOR ALIEN LIFE

Jupiter's icy moon Europa likely has a salty ocean beneath its frozen crust. It's possibly twice the volume of all Earth's oceans combined. Tidal heating from Jupiter's gravity may keep it warm enough for microbial life to exist.

THERE'S A PLANET THAT RAINS MOLTEN GLASS

Exoplanet HD 189733b is a scorching gas giant 63 light-years away. Winds whip through its atmosphere at 5,400 mph, and it rains molten glass—*sideways*—due to those high speeds and temperatures.

OUR MILKY WAY IS ON A COLLISION COURSE WITH ANOTHER GALAXY

In about four billion years, the Milky Way will collide with its neighbor, the Andromeda Galaxy. The result? A massive new galaxy often dubbed Milkomeda. But don't worry. Stars are so spread out that most won't actually crash.

TITAN HAS LAKES AND RAINSTORMS OF LIQUID METHANE

Saturn's moon Titan is the only moon in the solar system with a thick atmosphere and surface liquid. But unlike Earth, its rivers and lakes are made of methane and ethane.

THERE'S A ROGUE PLANET DRIFTING ALONE

Not all planets orbit stars. Some, like PSO J318.5-22, wander the galaxy alone. This rogue planet is a massive, Jupiter-sized body floating freely through space.

LIFE ABOARD THE INTERNATIONAL SPACE STATION

ASTRONAUTS GROW TALLER IN SPACE

In microgravity, astronauts can grow up to two inches taller because their spines elongate without the constant compression of Earth's gravity. Unfortunately, they shrink back down after returning to Earth.

SWEAT DOESN'T DRIP

On the International Space Station (ISS), there's no gravity to pull sweat off skin. Instead, it clings to the body in floating blobs, which astronauts have to wipe off with towels to avoid sticky discomfort.

ASTRONAUTS SEE 16 SUNRISES AND SUNSETS PER DAY

The ISS orbits Earth about every 90 minutes, which means astronauts witness a sunrise or sunset roughly every 45 minutes. That's 16 sunrises and 16 sunsets in just one Earth day!

NO BREAD ALLOWED!

Crumbs are a big problem in zero gravity, so bread is banned aboard the ISS. Astronauts eat things like tortillas instead. Every meal is sealed, labeled, and designed for both safety and nutrition.

SPACE MISSIONS THAT WENT WRONG OR GOT WEIRD

APOLLO 12 WAS STRUCK BY LIGHTNING — TWICE

Just 36 seconds after liftoff in 1969, Apollo 12 was hit by lightning, then struck again 16 seconds later. The electrical strikes knocked out critical systems, but a quick-thinking engineer on the ground suggested flipping a single switch that miraculously brought the mission back online.

NASA ACCIDENTALLY ERASED THE ORIGINAL MOON LANDING TAPES

In the 1980s, NASA accidentally recorded over the original broadcast-quality video of the Apollo 11 moon landing during a data tape shortage. The videos we see today are restored broadcast copies, not the original footage.

COSMONAUTS TOOK SHOTGUNS INTO SPACE

Soviet cosmonauts on early missions like Voskhod were equipped with triple-barreled pistols (TP-82s). They needed weapons in case they landed off course in Siberia, where they'd have to defend themselves from wolves and bears while waiting for rescue.

A NASA ASTRONAUT DROVE ACROSS THE U.S. IN A DIAPER

In 2007, astronaut Lisa Nowak was arrested after a bizarre cross-country trip to confront a romantic rival. She drove 900 miles while wearing adult diapers to avoid restroom stops. She brought pepper spray, a wig, and a BB gun. The incident became tabloid and courtroom drama.

GEMINI 8 NEARLY SPUN OUT OF CONTROL

In 1966, Gemini 8 began spinning violently at one revolution per second due to a stuck thruster. Commander Neil Armstrong manually disabled the main system and used backup thrusters to regain control, saving the mission and their lives.

WILD THEORIES ABOUT THE UNIVERSE

WE LIVE IN A MULTIVERSE

The multiverse theory posits that our universe is just one of an infinite number of universes, each with different physical laws, histories, and versions of ourselves. In some, you may be a pirate; in others, Earth might not exist at all.

THE UNIVERSE MIGHT BE INSIDE A BLACK HOLE

One mind-bending theory suggests that our entire universe exists inside a black hole located in a larger universe. Black holes, according to general relativity, can contain their own spacetime and laws of physics. This theory could explain our own universe's mysterious expansion.

DARK MATTER MIGHT BE SHADOW PARTICLES FROM ANOTHER REALM

Dark matter makes up 85% of the matter in the universe, yet we can't see it or interact with it directly. Some theories say it might be particles from a parallel universe that coexists with ours but only interacts through gravity.

THE BIG BANG MIGHT NOT HAVE BEEN THE BEGINNING

Some scientists theorize that the Big Bang wasn't the beginning but rather a "bounce" in an eternal cycle of expansion and contraction. This cyclic model suggests that the universe has existed forever in repeating patterns of birth and rebirth.

MIND-BENDING SPACE PHENOMENA

WHITE HOLES, THE THEORETICAL OPPOSITES OF BLACK HOLES

A white hole is a hypothetical region in space-time that can't be entered from the outside, but matter and light can escape from it. They're the reverse of black holes; instead of sucking things in, they constantly expel them. Some scientists theorize that white holes could be connected to black holes via wormholes.

QUASARS: BRIGHTER THAN A TRILLION SUNS

Quasars are the ultra-luminous cores of young galaxies. They're powered by supermassive black holes devouring matter and can emit more energy than entire galaxies. They're so distant and bright that we see light from them billions of years in the past.

GAMMA-RAY BURSTS ARE THE UNIVERSE'S MOST VIOLENT EXPLOSIONS

These cosmic flashes last only seconds but release more energy than our sun will emit over its entire 10-billion-year life. If one occurred nearby, it could strip Earth's atmosphere. Luckily, they usually happen in distant galaxies.

THE BOÖTES VOID

The Boötes Void is a massive, nearly empty region of space that spans about 330 million light-years across. Sometimes called the "Great Nothing," it contains very few galaxies compared to the rest of the universe. Its eerie emptiness challenges our understanding of cosmic structure and formation.

FAMOUS ASTRONAUTS AND THEIR FUN SPACE STORIES

JOHN YOUNG SNUCK A CORNED BEEF SANDWICH INTO SPACE

During the 1965 Gemini 3 mission, astronaut John Young secretly brought a corned beef sandwich aboard. It caused controversy because crumbs could be hazardous in microgravity, but it also made history as the first deli meat in space.

CHRIS HADFIELD PLAYED MUSIC IN SPACE

Canadian astronaut Chris Hadfield went viral for his cover of David Bowie's *Space Oddity*, recorded aboard the ISS. He was the first person to record a music video in space, and he often entertained Earth with videos about daily life aboard the station.

BUZZ ALDRIN TOOK COMMUNION ON THE MOON

Before stepping onto the lunar surface, Buzz Aldrin took a private communion using bread and wine he brought from Earth. It was a deeply personal moment he kept mostly quiet about at the time, making it the first religious sacrament performed on the moon.

YURI GAGARIN'S PREFLIGHT RITUALS

Before becoming the first human in space in 1961, Yuri Gagarin stopped the bus on the way to the launch pad to pee on the rear tire. This spontaneous moment became a weird but beloved tradition among Russian cosmonauts to this day.

THE ONGOING SEARCH FOR EXTRATERRESTIAL LIFE

THE JAMES WEBB SPACE TELESCOPE IS ANALYZING ALIEN ATMOSPHERES

NASA's James Webb Space Telescope is now being used to examine planets outside our solar system for gases like oxygen, methane, and carbon dioxide that might suggest the presence of life. The telescope studies light filtering through the planets' atmospheres to identify potentially habitable worlds.

THE "WOW! SIGNAL" STILL BAFFLES SCIENTISTS

In 1977, a radio telescope in Ohio detected a mysterious 72-second radio burst from space that's now known as the "Wow! signal." To this day, it remains one of the best candidates for a potential alien transmission. It has never been repeated or explained.

MARS MISSIONS ARE ZEROING IN ON ANCIENT RIVERBEDS

Rovers like Perseverance are exploring Martian river deltas in the Jezero Crater, a location scientists believe was once a lakebed. The goal? To find microfossils or organic molecules — direct evidence that life may have once existed on the Red Planet.

EUROPA AND ENCELADUS MAY CONTAIN ALIEN LIFE IN THEIR OCEANS

Jupiter's moon Europa and Saturn's moon Enceladus have subsurface oceans beneath their icy crusts. These oceans may contain the right chemical ingredients for life. NASA is planning missions (like the Europa Clipper probe) to fly through water plumes that erupt from the surface and look for organic compounds.

SETI IS NOW USING AI TO SCAN THE STARS

SETI, the Search for Extraterrestrial Intelligence, has started employing machine learning algorithms to sift through massive amounts of radio signal data. AI helps flag unusual or non-random patterns that might originate from an intelligent source rather than from natural phenomena or human interference.

CHAPTER ELEVEN: FOOD AND DRINK

SURPRISING FOOD ORIGINS AND NAMING HISTORIES

KETCHUP STARTED AS FERMENTED FISH SAUCE

The term *ketchup* comes from the Hokkien Chinese word *kê-tsiap*, a fermented fish brine. When British traders encountered it in Southeast Asia in the 17th century, they brought the concept back but eventually replaced the fish with tomatoes, sugar, and vinegar.

CROISSANTS ARE AUSTRIAN, NOT FRENCH

Despite being a symbol of French culture, the croissant actually originates from Vienna, Austria. It was inspired by the Kipferl, a crescent-shaped pastry that dates back centuries. The croissant as we know it was popularized in France when Austrian baker August Zang opened a Viennese bakery in Paris in the 1830s.

PEANUTS AREN'T NUTS; THEY'RE LEGUMES

Peanuts grow underground, while true nuts like almonds or walnuts grow on trees. Botanically speaking, peanuts are legumes, making them related to beans and lentils. The term *nut* in the name is a misnomer that stuck due to their similar texture and culinary uses.

WORCESTERSHIRE SAUCE WAS AN ACCIDENTAL DISCOVERY

Worcestershire sauce originated in England when chemists John Wheeley Lea and William Henry Perrins attempted to recreate a sauce from India. Their initial concoction was so awful that they shelved it, only to rediscover it months later after it had fermented.

The aged version had a complex, savory flavor, and the now-famous condiment was born.

SANDWICHES WERE NAMED AFTER A GAMBLING HABIT

The sandwich is named after 18th-century English nobleman John Montagu, the 4th Earl of Sandwich. Legend says he didn't want to leave the gambling table to eat, so he requested his meat be served between two slices of bread. That way, he could eat with one hand and keep playing with the other.

STRANGE INGREDIENTS USED IN EVERYDAY SNACKS

CRUSHED BUGS IN RED CANDY

That bright red color in candies like red Skittles, Starburst, and some fruit yogurts often comes from cochineal, a dye made of crushed insects. The dye is also labeled as *carmine* or *E120*.

BEAVER GLANDS IN VANILLA AND RASPBERRY FLAVORING

Castoreum, a substance made from the scent glands of beavers, has historically been used as a natural flavoring in vanilla and raspberry foods. While rare today due to cost and regulation, it was once common in ice cream and candy. It may still appear under the vague label *natural flavoring*.

SHELLAC IN SHINY CANDY COATING

That glossy finish on jelly beans and shiny chocolate candies? It might come from shellac, a resin secreted by the female lac bug. Shellac is also used in wood varnish.

HUMAN HAIR OR DUCK FEATHERS IN BREAD

Commercial bread and bakery products often contain L-cysteine, an amino acid used to improve dough texture. This additive can be sourced from human hair, duck feathers, or hog bristles, but many manufacturers now opt for synthetic or plant-based versions.

WOOD PULP IN SHREDDED CHEESE AND ICE CREAM

Ever see *cellulose* listed in ingredients? That's wood pulp. It's commonly added to shredded cheese to prevent clumping and sometimes used in low-fat ice creams as a thickener. It's technically edible fiber, but it's… literally wood.

REGIONAL DISHES THAT SOUND (OR TASTE) BIZARRE

HÁKARI: FERMENTED SHARK

This Icelandic dish consists of Greenland shark meat that's been buried underground to rot for months, then hung to dry. The result smells strongly of ammonia and has a chewy texture. Even celebrity chef Anthony Bourdain called it "the single worst, most disgusting and terrible-tasting thing." Locals often chase it with a shot of Brennivín (a strong schnapps).

CASU MARZU, THE MAGGOT CHEESE

This infamous rotten cheese is made from Pecorino and deliberately infested with live cheese fly larvae. The maggots help ferment the cheese, making it soft and intense in flavor. It's often eaten with the larvae still wriggling inside. Although technically illegal in the European Union, it remains a cultural tradition in Sardinia, Italy.

ROCKY MOUNTAIN OYSTERS

Despite the name, Rocky Mountain oysters aren't seafood; they're deep-fried bull testicles. This delicacy is commonly served at ranch festivals in parts of the American West, especially in Colorado. Crunchy, chewy, and often paired with cocktail sauce, they're a cowboy classic.

BALUT: DUCK EMBRYO EGG

A beloved Filipino street snack, balut is a fertilized duck egg with a partially developed embryo inside. The egg is boiled and eaten whole, including the bones and feathers. Often seasoned with vinegar or salt, it's considered a high-protein delicacy and a rite of passage for adventurous eaters.

STARGAZY PIE

This British pie from Cornwall features whole sardines or pilchards—heads and all—baked into a pie crust with their little faces sticking out. It was traditionally made to celebrate fishermen surviving a storm and bringing home a catch. The fish heads looking skyward, or stargazing, supposedly let the oils run back into the pie.

FOOD INVENTIONS THAT HAPPENED BY ACCIDENT

POTATO CHIPS WERE A REVENGE PRANK

In 1853, chef George Crum in New York created what we now know as potato chips out of spite. A customer kept complaining that his fried potatoes were too thick, so Crum sliced them paper-thin, over-fried them until crispy, and drowned them in salt. Instead of being angry, the customer loved them, and chips were born.

POPSICLES WERE INVENTED BY AN 11-YEAR-OLD

In 1905, a boy named Frank Epperson accidentally left a cup of soda powder and water outside overnight with a stir stick still in it. It froze overnight, and when he tasted it the next morning, the Popsicle was born. He called it the Epsicle for years before changing the name.

CHOCOLATE CHIP COOKIES CAME FROM A BAKING SHORTCUT

In the 1930s, Ruth Wakefield of the Toll House Inn in Massachusetts ran out of baking chocolate. She substituted broken Nestlé semi-sweet chocolate chunks, expecting them to melt into the dough. They didn't, and instead, she created the world's first chocolate chip cookie.

CORN FLAKES WERE A SANITARIUM EXPERIMENT GONE WRONG

In the late 1800s, John and Will Kellogg were trying to create a bland, healthy food for patients at their health spa. They

110

accidentally left boiled wheat sitting out too long, and when they rolled it out, it flaked. They repeated the process with corn and made corn flakes, launching the breakfast cereal industry.

CHEWING GUM BECAME POPULAR THANKS TO A FAILED INVENTION

Thomas Adams was trying to turn chicle, a sap from the sapodilla tree, into rubber. That didn't work, but he discovered it made a great chewable substance. He added flavor and started selling it as Adams New York Chewing Gum, helping to spark the modern gum industry.

RECORD-BREAKING MEALS AND EATERS

JOEY CHESTNUT ATE 76 HOT DOGS IN 10 MINUTES

At the 2021 Nathan's Hot Dog Eating Contest, competitive eating legend Joey Chestnut set a world record by eating 76 hot dogs and buns in just 10 minutes. That's roughly 22,800 calories, more than most people eat in a week.

THE LONGEST SANDWICH WAS OVER TWO MILES LONG

In 2004, Italy built the world's longest sandwich, measuring 7,913 feet. That's over two miles of sandwich! It featured salami and cheese, and it took a team of chefs working in relay to assemble and serve it.

THE LARGEST CHEESEBURGER WEIGHED OVER A TON

A restaurant in Mallie, Kentucky, made the heaviest cheeseburger ever in 2017 after four years of planning. It weighed a staggering 2,014 pounds and included 300 pounds of cheese, 200 pounds of bacon, and an 8-foot-wide bun!

TAKERU KOBAYASHI ATE 69 HOT DOGS DESPITE WEIGHING JUST 128 POUNDS

Japanese competitive eater Takeru Kobayashi revolutionized eating contests in the early 2000s. At only 5'8" and 128 pounds, he once ate 69 hot dogs in 10 minutes using his own "Solomon technique" (breaking the dogs in half). He turned competitive eating into a science—and a sport.

DANGEROUS FOODS PEOPLE ACTUALLY EAT

FUGU: THE POISONOUS PUFFERFISH

Fugu is a luxury Japanese dish made from pufferfish. It contains tetrodotoxin, a neurotoxin 1,200 times more deadly than cyanide. Chefs must train for years and be licensed to serve it since one small mistake in preparation can be fatal. Despite the danger, it's a revered winter delicacy.

ACKEE: A FRUIT THAT CAN KILL

Ackee, Jamaica's national fruit, is only safe to eat when fully ripe and naturally opened. If consumed too early or improperly prepared, it can cause Jamaican vomiting sickness due to the toxin hypoglycin A. It's delicious sautéed with saltfish but deadly if done wrong.

SANNAKJI: LIVE OCTOPUS TENTACLES

Sannakji is a Korean dish made from live baby octopus, served with its tentacles still writhing on the plate. The danger? The suckers can still grip and stick to your throat, leading to choking deaths every year. Chewing thoroughly is mandatory and part of the thrill.

NUTMEG: A SPICE WITH A TOXIC TWIST

In small amounts, nutmeg is harmless. But in large doses of two to three tablespoons, it can cause hallucinations, seizures, nausea, and even death. The side effects are due to the compound myristicin. Nutmeg has been used recreationally, but experts warn against it for very good reasons.

HISTORICAL USES OF FOOD AS MEDICINE

GARLIC WAS ANCIENT EGYPT'S PERFORMANCE ENHANCER

Ancient Egyptians fed garlic to laborers and pyramid builders as a strength and endurance booster. It was believed to ward off disease and evil spirits. Today, garlic is still praised for its antibacterial and heart-healthy properties.

HONEY WAS USED AS AN ANTISEPTIC IN WOUND CARE

The Egyptians, Greeks, and Romans used raw honey to treat wounds because of its natural antibacterial properties. It also helped preserve the body of Alexander the Great, whose corpse was allegedly transported in a honey-filled coffin to prevent decay.

WILLOW BARK: THE ORIGINAL ASPIRIN

Used in Ancient Greece and China, willow bark tea was prescribed for fever and pain. It contains salicin, a compound later synthesized into aspirin in the 1800s. Hippocrates recommended it as early as 400 BCE for childbirth pain.

TURMERIC WAS A GOLDEN CURE IN AYURVEDA

In Indian Ayurvedic medicine, turmeric has been used for over 4,000 years to treat conditions such as inflammation and digestive issues. Its active compound, curcumin, has recently been confirmed to have anti-inflammatory and antioxidant effects in modern studies.

WINE WAS PRESCRIBED FOR EVERYTHING IN ANCIENT ROME

Roman physicians like Galen regularly prescribed wine as a digestive aid, antiseptic, and sleep remedy. It was mixed with herbs and used to treat everything from melancholy to epilepsy. While we now understand alcohol's risks, red wine in moderation is still studied for its potential heart benefits.

CHAPTER TWELVE: INVENTIONS AND DISCOVERIES

INVENTORS WHO WERE OVERLOOKED OR ROBBED

NIKOLA TESLA: THE FORGOTTEN FATHER OF ELECTRICITY

Although Tesla's inventions in alternating current electricity revolutionized power systems, Thomas Edison often got the spotlight and financial backing. Tesla died nearly penniless, while Edison became a household name. Tesla's contributions were only widely celebrated posthumously.

ROSALIND FRANKLIN: DNA'S UNSUNG HERO

Rosalind Franklin's X-ray crystallography images were critical to discovering the double helix structure of DNA. However, James D. Watson and Francis Crick received most of the credit and the Nobel Prize, while Franklin died young and unrecognized during her lifetime for this groundbreaking work.

PHILO FARNSWORTH: TV'S LOST PIONEER

Philo Farnsworth invented the first fully electronic television system in the 1920s, but RCA and David Sarnoff fought him in court and dominated the market. Farnsworth's patents were eventually bought out, and his name became less known compared to other industry giants.

HENRIETTA LEAVITT: THE WOMAN WHO MAPPED THE STARS

Leavitt discovered the relationship between a Cepheid star's brightness and its distance, an essential step for measuring the universe's size. Her work was overshadowed by male astronomers, and she never received the recognition she deserved during her lifetime.

LÁSZLÓ BÍRÓ: THE PEN THAT CHANGED THE WORLD

Although Bíró invented the modern ballpoint pen, he initially struggled to patent it and bring it to market. It was later mass-produced by others who made millions while Bíró lived modestly.

INNOVATIONS INSPIRED BY NATURE

VELCRO WAS INSPIRED BY BURRS

In 1941, Swiss engineer George de Mestral noticed that burrs clung stubbornly to his dog's coat after a hike. Under a microscope, he saw tiny hooks that latched onto loops in fabric and fur. His observations led to the invention of Velcro, which is now used in everything from sneakers to NASA gear.

BULLET TRAINS MODELED AFTER KINGFISHERS

Japan's original bullet trains created sonic booms when exiting tunnels. The redesign was inspired by the kingfisher bird, whose sleek beak allows it to dive into water with barely a splash. The new design reduced noise and increased speed and energy efficiency.

TERMITE MOUNDS INSPIRED ECO-FRIENDLY ARCHITECTURE

Termite mounds in Africa maintain constant internal temperatures despite external heat. Architects studied these structures to design buildings like the Eastgate Centre in Zimbabwe, which uses natural ventilation and no air conditioning, cutting energy use by 90%.

LOTUS LEAVES INSPIRED SELF-CLEANING SURFACES

The lotus leaf has microscopic bumps that repel water and dirt. This phenomenon inspired self-cleaning materials. The water droplets roll off, taking dirt with them, just like on the lotus.

GECKO FEET LED TO SUPER-GRIP ADHESIVES

Geckos can climb smooth vertical surfaces thanks to microscopic hairs on their feet called setae. These hairs exploit van der Waals forces to cling tightly without glue. Dry adhesives mimic the effect for use in robotics and rescue missions.

ANCIENT INVENTIONS WE STILL USE TODAY

ROMAN CONCRETE

The Romans invented a type of concrete over 2,000 years ago that's more durable than many modern versions. Their secret? Mixing volcanic ash, lime, and seawater. Ancient Roman harbors made with this formula still stand today, outlasting some modern ports.

THE ALARM CLOCK

The ancient Greek philosopher Plato reportedly used a water-based alarm clock from 428 to 348 BCE to signal the start of his lectures at dawn. Later, Greek and Roman inventors improved the design. The concept of a device to wake us up has been in use ever since.

SURGERY TOOLS

The *Sushruta Samhita*, a Sanskrit medical text from India (circa 600 BCE), describes over 100 surgical instruments. Many of these resemble modern-day scalpels, forceps, and scissors. Roman surgical kits found in Pompeii look remarkably similar to tools still used today.

THE FLUSH TOILET

Around 2,800 BCE, the Minoans of Crete had flushing toilets and sophisticated plumbing systems that used gravity and water pressure. The Romans later built on this idea with public toilets and underground sewage systems, precursors to today's sanitation infrastructure.

THE POSTAL SYSTEM

The Persian Empire under Cyrus the Great (circa 550 BCE) created one of the first organized postal systems, using mounted couriers and relay stations. The motto of the U.S. Postal Service—"neither snow nor rain..."—is derived from a description of this system by Herodotus.

THE ROLE OF WAR IN SPEEDING UP INNOVATION

WWII GAVE US THE MODERN COMPUTER

British mathematician Alan Turing developed an early computer called the Bombe to crack Nazi codes. This cryptographic breakthrough laid the foundation for modern computing and inspired future generations of tech innovation, including the development of the internet.

CANNED FOOD WAS INVENTED FOR NAPOLEON'S ARMY

In the early 1800s, Napoleon offered a reward for a food preservation method to support his troops. Nicolas Appert won by creating a system of sealing food in glass jars (a precursor to modern canning). It revolutionized military and civilian food storage.

DUCT TAPE WAS MADE TO SEAL AMMO BOXES

During WWII, Johnson & Johnson developed a waterproof adhesive tape for the U.S. military to seal ammunition boxes. Troops quickly discovered its versatility and used it for repairs in the field. After the war, it became the beloved duct tape we know today.

JET ENGINES TOOK FLIGHT BECAUSE OF WAR

While jet propulsion was theorized earlier, World War II drove the rapid development of jet engines. Both the Allied and Axis powers raced to produce faster, more efficient aircraft. This war-fueled leap paved the way for commercial air travel after the war.

CHAPTER THIRTEEN: LANGUAGE AND WORDS

THE WEIRD ORIGINS
OF EVERYDAY EXPRESSIONS

THE LONGEST WORD IN ENGLISH

The full chemical name for titin, the largest known protein, is technically the longest word in English. With 189,819 letters, it takes over three and a half hours to pronounce! Although not in dictionaries, it's often cited to demonstrate how scientific terms can stretch the boundaries of language.

THE WORD *QUARANTINE* COMES FROM THE ITALIAN WORDS FOR "40 DAYS"

During the Black Death in the 14th century, ships arriving in Venice were required to wait *quaranta giorni* (40 days) before passengers could disembark. This gave rise to the term *quarantine*.

GOODBYE IS A CONTRACTION OF A RELIGIOUS PHRASE

Goodbye originally came from the phrase "God be with ye." Over time, it was shortened and phonetically altered into the word we use today. It's one of many examples of how everyday words hide deep historical roots.

SOME LANGUAGES HAVE NO WORD FOR LEFT OR RIGHT

Certain Indigenous Australian languages like Guugu Yimithirr use absolute directions such as north and east instead of relative terms. Speakers develop a constant sense of cardinal orientation, essentially living with a built-in compass!

THE WORD *ROBOT* COMES FROM A 1920 PLAY

The word *robot* was first used in the play *Rossum's Universal Robots* by Czech writer Karel Čapek. It comes from *robota*, meaning "forced labor or drudgery" in Czech. Today, it defines a whole field of science and fiction.

WORDS THAT DON'T EXIST IN ENGLISH

JAPANESE: *TSUNDOKU*

Tsundoku describes the act of acquiring books and letting them pile up unread. English speakers might say "book hoarding," but *tsundoku* captures the well-intentioned neglect of to-be-read stacks, a feeling many readers know all too well.

GERMAN: *WALDEINSAMKEIT*

This beautiful term refers to the feeling of being alone in the woods in a peaceful and reflective way. It evokes a mixture of serenity and introspection that English only circles around with poetic phrases.

TAGALOG: *KILIG*

Kilig is the fluttery feeling of excitement you get when something romantic happens, like a crush smiling at you. It's not just butterflies in your stomach; it's the giddy thrill of love in its earliest sparks.

PORTUGUESE: *SAUDADE*

Saudade describes a deep, emotional state of nostalgic longing for something or someone that may never return. It's a bittersweet yearning for the past, a place, a person, or even a version of yourself.

SWEDISH: *GÖKOTTA*

This term means to wake up early and listen to the first birds singing. A poetic tradition turned into a word, gökotta suggests a peaceful, mindful start to the day. It's something many of us do but don't have a name for in other languages.

TONGUE TWISTERS AND TRICKY PHRASES

THE WORLD'S HARDEST TONGUE TWISTER

According to a 2013 MIT study, the phrase "pad kid poured curd pulled cod" is one of the hardest tongue twisters in English. It was so difficult that some test subjects couldn't even finish saying it!

TONGUE TWISTERS TRAIN YOUR BRAIN

Tongue twisters aren't just silly fun; they're used by actors, singers, and speech therapists to warm up the vocal cords and strengthen pronunciation. They challenge the brain's phonological loop, which helps process sound patterns.

"SHE SELLS SEASHELLS" IS BASED ON A REAL WOMAN

The classic tongue twister "She sells seashells by the seashore" is believed to be inspired by Mary Anning, a 19th-century fossil collector and paleontologist. She sold ancient shells and bones along the English coast.

124

SOME TONGUE TWISTERS ARE DESIGNED TO SOUND WRONG

Many tongue twisters use phonetic proximity—words that sound nearly the same but aren't—to trick your brain into making mistakes. They highlight how our speech planning can "crash" under pressure.

OTHER LANGUAGES HAVE THEIR OWN TONGUE TWISTERS

English doesn't have a monopoly on linguistic chaos. In Mandarin, there's a famous one that uses only the syllable *shi* with different tones! In Czech, "Strč prst skrz krk" means "stick your finger through your throat." It forms a full sentence without a single vowel.

ETYMOLOGY OF FUNNY OR GROSS WORDS

FART IS ONE OF THE OLDEST WORDS IN THE ENGLISH LANGUAGE

The word *fart* dates back over 1,000 years to the Old English term *feortan*, meaning "to break wind." Its roots are Proto-Indo-European, making it a linguistic fossil that's still alive and tooting.

BOOGER LIKELY COMES FROM *BOGEY*

This childhood gross-out word may trace back to the Middle English term *bogge*, meaning "ghost" or "goblin." It likely shares a root with *bogeyman*. The idea of something nasty hiding inside your nose fits the creepy lineage perfectly.

PUKE MIGHT BE SHAKESPEAREAN

The word *puke* was popularized by Shakespeare in *As You Like It* (1599) when he wrote about different stages of life. One of the lines says, "The infant, mewling and puking in the nurse's arms."

DORK ORIGINALLY MEANT "WHALE PENIS"

This slang insult began as 1960s American teenage jargon, but some etymologists point to older nautical slang where *dork* referred to a whale's reproductive organ. Over time, it lost its anatomical meaning and became a go-to jab for awkwardness.

NINCOMPOOP HAS MYSTERIOUS ORIGINS

While it sounds like sheer nonsense, one theory is that *nincompoop* may come from the Latin *non compos mentis*, meaning "not of sound mind." Others believe it came from a legal term or even a misheard phrase. Either way, it morphed into a goofy insult for someone clueless.

PUNCTUATION AND LANGUAGE EVOLUTION

THE QUESTION MARK MAY HAVE EVOLVED FROM LATIN

One theory suggests that the question mark originated from the Latin word *quaestio*, meaning "question." When written, it was abbreviated to *Qo*. Scribes eventually stacked the *Q* over the *o*, which over centuries evolved into the familiar hook-and-dot symbol we use today.

THE AMPERSAND WAS ONCE THE 27TH LETTER OF THE ALPHABET

In the 1800s, children reciting the alphabet would end with "X, Y, Z, and per se and." That last phrase, said quickly, became *ampersand*. It wasn't punctuation; it was a full-blown letter!

OLD ENGLISH HAD LETTERS WE DON'T USE ANYMORE

Early English used letters like *thorn* (þ) and *eth* (ð) to represent *th* sounds. If you've ever seen a store called Ye Olde Shoppe, that Y was actually a misprinted þ. It was never meant to be pronounced "Yee"!

DOUBLE SPACES AFTER PERIODS CAME FROM TYPEWRITERS

Using an extra space after a period was standard when monospaced fonts (like on typewriters) were common. With proportional fonts used in digital typography, it's no longer needed, but the habit lives on in many people trained before the 2000s.

CONCLUSION

The world is full of weird and surprising facts, and now you know a whole lot more of them! From animals with wild abilities to scientists who made life-changing discoveries by accident, we've explored just how strange and amazing our universe really is. Every chapter in this book shows that there's always more to learn.

Maybe you were surprised that octopuses have blue blood or that your body glows in the dark. Maybe you were amazed to learn about people who can run for hours without sleep. Or maybe you just had fun discovering odd traditions and wacky inventions.

The coolest part? These facts help us think differently about science, technology, and ourselves. They remind us that curiosity is powerful. Asking "Why?" or "How?" can lead to some of the biggest discoveries in human history.

Don't stop here. Keep exploring and asking questions. Look deeper into the things that interest you. Being curious is one of the smartest things you can be.

Thanks for reading. Stay curious, stay sharp, and never stop wondering about the world around you.